MASTERING TECHNOLOGY

A Management Framework for Getting Results

Rod F. Monger

THE FREE PRESS
A Division of Macmillan, Inc.
NEW YORK

Collier Macmillan Publishers
LONDON

The Free Press
A Division of Macmillan, Inc.
866 Third Avenue, New York, N.Y. 10022

Collier Macmillan Canada, Inc.

Printed in the United States of America

printing number
1 2 3 4 5 6 7 8 9 10

Library of Congress Cataloging-in-Publication Data

Monger, Rod F.
Mastering technology.

Bibliography: p.
1. Technological innovations—Management. 2. Technology—Management. 3. Information technology—Management. I. Title.
HD45.M66 1988 658.5′14 88-2810
ISBN 0-02-921680-X

We acknowledge the following for text quoted on pp. 1, 2, and 3 of this volume: J. J. Servan-Schreiber, excerpted from *The American Challenge*. Copyright © 1967 Editions Denoël as *Le Défi Américain*. English translation copyright © 1968, 1969 Atheneum House, Inc. Reprinted with the permission of Atheneum Publishers, an imprint of Macmillan Publishing Company.

Contents

List of Figures and Tables

Figures

Tables

Preface

Over the past couple of decades an important shift in the global economic balance of power has been occurring. Once in a lofty position of world dominance, the United States has lost its considerable lead over many of its major trading partners. Japan has risen from the ashes of World War II to become a supercompetitor, challenging the United States by winning large market shares in industries such as automobiles and electronics. Other Pacific Rim nations such as Taiwan and South Korea are following Japan's lead, intensifying international competition even more. These new entrants have changed the basis of competition. Operating under different financial, political, and social constraints, they pursue markets with an emphasis on quality, customer satisfaction, and low-cost production.

American business has been challenged to rethink its entire approach to competition. Previously, managers of large companies thought about competition in terms of other American companies in the industry, companies that played by the same rules. Now, the emphasis is on restructuring businesses to be competitive with both American and foreign firms. At stage center in this new competitive arena are information technologies. These technologies, including central data processing, telecommunications, office automation, and computer-integrated manufacturing, are critical to the economic future of the United States. Yet for many businesses, information technologies have not met expectations. The impact on productivity has been disappointing and computer assets are underutilized. Why?

Answering this question is the mission I have set out to accomplish in this book. Part of the answer is that we think about technological change in revolutionary terms, and this interpretation of events undermines management practices. A de facto philosophy has evolved which says, in effect, that because technology is always changing, our approach to management must change as well, and that management methods of the past offer little guidance for managing the future. This interpretation is wrong. It encourages management to remain remote and uninvolved with technology and the process of change it creates within their organizations, or to delegate the responsibility for these changes to subordinates who are unqualified to make the decisions that matter. Consequently, in firms where technology has an important role to play in business competitiveness, major opportunities for strategic advantage are overlooked. When managers do become involved, their time and resources are often squandered on inconsequential technical matters.

A more useful view suggests that after more than three decades of living with computers, we should be able to examine our experiences and better understand what management's role and responsibilities are with respect to technology. What have we done successfully? What have we done incorrectly? What in the way of management practices has really changed? Most importantly, what has remained constant and enduring throughout? In essence, we are asking what basic principles can be distilled from our experiences that will help us in *mastering technology?*

A portion of the answer lies in the context in which technology must be managed. For reasons largely historical in nature, American managers now find themselves in an atmosphere that is hostile and corrosive to effective decision making about technology. Many of the economic signals given to managers by current accounting and financial conventions are wrong. Managers often instinctively know the answers to many technology-related questions. For example, white-collar productivity will obviously improve with automation. Yet managers strive in vain looking for ways to cost justify this belief. The accounting profession and others are even beginning to consider new methodologies for measuring these presumed benefits. But as we shall see in due course, the problem does not lie with the technology or with measurement methodologies per se. Indeed, the assumptions that underlie decision making are false. We cannot expect to improve the performance of technology in American business, nor, for that matter, the performance of American business in international competition, until these contextual issues are addressed.

To satisfy ourselves, we must look beyond organizational boundaries. What are the factors in the general business environment that are barriers to reasoned decision making? What must be done to remove the impediments to effective competition by the United States in the evolving global economy? How will the removal of these factors improve management of information technologies?

My approach to answering these questions has been to look at the history of traditional management thought and practice and to apply it to technology-related issues in an attempt to identify themes that are repeated each time a new wave of innovations comes along. Based on these themes, an Integrated Technology Management Framework is presented and issues relevant to the principles embodied in this Framework explored. Although every attempt has been made to make the discussion relevant to current technology, it is hoped that the Framework and its principles will endure and provide guidance long after the current wave of computer hardware and software is obsolete.

This book has been written to avoid blinding readers with technical detail. "Heat-seekers," those people who are the first on their block to savor each technical innovation, will be keenly disappointed. They may, however, find the contents liberating. The primary audience is the senior managers, those responsible for overall organizational leadership. For them I have tried to provide a practical guide, avoiding the academic approach. As Mark Twain once said "If it were a record of a solemn scientific exploration, it would have about it that gravity, that profundity, and that impressive incomprehensibility which are so proper to works of that kind, and withal so attractive." But, alas, it is not. I hope that managers in general, corporate directors, academicians, business students, and consultants will find its contents useful as well.

The material for this volume was gathered from a broad and diverse range of sources. I relied on numerous articles and books from the management literature which are, of course, listed as references. I also tapped the experience and judgment of colleagues in business and academia as well as my own background as educator, consultant, and information systems practitioner. Some of the material was drawn from original research I conducted as a member of the faculty at Fordham University's Graduate School of Business Administration. This research was supported, in part, by a grant from Fordham.

I am especially grateful to Joseph T. Brophy and Trav Waltrip at The Travelers Companies in Hartford, Connecticut; Tom Murrin at

Westinghouse Corporation in Pittsburgh; Edward Nyce at Manufacturers Hanover Trust Company in New York; Antoinette La Belle at Kidder Peabody (formerly at Manufacturers Hanover Trust Company); a number of folks at IBM, especially at the Lexington, Kentucky plant; and others in these companies and elsewhere, too numerous to mention, for their time and the information and insights they shared.

Claudette Bailey made an enormous contribution by seeing that the logistics of producing the manuscript, which I had woefully underestimated, proceeded smoothly.

While writing this book, I was also deeply engaged in developing a new academic program in information and communications systems management at Fordham University's Graduate School of Business Administration. In conjunction with this effort, a small band of very bright and dedicated students helped me build The Technology Center, a place for innovation, experimentation, and learning with technology. The loyalty and enthusiasm of these students—Ray Menendez, Mike Tepedino, Chris Unrath, and John Walsh—sustained me throughout. May the road rise to meet you, fellas.

1

The Missing Link: Technology Management

The Crisis in Managing Technology

As recently as twenty years ago, the United States was regarded as the undisputed leader in technology innovation and development and, so it was assumed, technology management. In 1967, a book entitled *The American Challenge* appeared which documented the dramatic achievements of the American economy. Among the torrent of statistics offered to prove the point, the author, Frenchman J. J. Servan-Schreiber, noted that the United States' share of world production in machinery was 70 percent, autos 76 percent, oil 73 percent, and chemicals 62 percent. Servan-Schreiber described this prosperity as follows:

> American industry produces twice the goods and services of all European industry combined—including both Britain and the Common Market—and two and a half times more than the Soviet Union, which has a greater population than the U.S. It produces a third of the total production of all other countries in the world. The Americans have achieved this with only 7 percent of the surface of the globe and 6 percent of its population.
>
> One third of all students in the world pursuing a higher educa-

tion are American. The number of students, as compared to the total population, is double that of any other country. For every 1,000 inhabitants there are 29 Americans studying at a university, 18 Russians, 10 Dutchmen and 10 Swedes.

All by themselves the Americans consume a third of the total world production of energy, and have one third of all the world's highways. Half the passenger miles flown every year are by American airlines. Two trucks of every five on the roads are American-made and American-based. Americans own three of every five automobiles in the world.

The combined profits of the ten biggest firms in France, Britain, and Germany (30 in all) are $2 billion. The profits of General Motors alone are $2.5 billion. To equal the profits of General Motors you would have to add the ten leading Japanese firms to the European total. These 40 firms employ 3.5 million people while General Motors employs 730,000—or about a fifth. (1967, pp. 49–50)

Servan-Schreiber wrote *The American Challenge* as a treatise to spur the European Economic Community into making reforms he felt were needed for economic expansion. He asserted that the United States was separated from the rest of the world by a "technological gap." This gap had been created, he believed, because of the emphasis on education and technological innovation. According to his sources, before 1929 most economic expansion was fueled by growth in both the labor force and invested capital. But after 1929 education and technology became more important. In Servan-Schreiber's words:

If the diffusion of education is the primary factor in economic development, the second is the "growth of knowledge"—the enrichment of education and its expansion to include adults, making the new technology available to them. (1967, p. 71)

Education and the interaction of education and technology increased productivity, and this fueled economic expansion. Servan-Schreiber dutifully supported his arguments with more statistics. Between 1930 and 1965, America had multiplied the sum spent on education tenfold. In 1965, 44 percent of university-age Americans were enrolled, compared to only 4 percent in 1900. "Growth of knowledge," which he measured in part by expenditures in research and development, increased more than a hundredfold between 1930 and 1964.

Servan-Schreiber also credited American management practices for the prolific success of the economy:

> Advanced technology and management skills have raised per capita production in the United States to a level of 40 percent above that of Sweden (next highest), 60 percent above Germany, 60 percent above France, and 80 percent above Britain. The driving force behind this power is American business. (1967, p. 50)

He noted that the U.S. gross national product had risen steadily for the previous six years, starting at 7.7 percent in 1961 and climbing to 9.5 percent in 1966. The reason for this performance, he claimed, was American management:

> Behind the success story of American industry lies the talent for accepting and mastering change. Technological advancement depends on virtuosity in management. Both are rooted in the dynamic vigor of American education.

By the time *The American Challenge* appeared, its assessment of the United States was already out of date. Unbeknownst to Servan-Schreiber the seeds of decline had already been sown. Had we been able to recognize the signs, we would have seen that early warnings had been posted that the economy was drifting off course. Servan-Schreiber unwittingly pinpointed the reason for the decline by quoting Kuan-Tzu, a Chinese poet from twenty-six centuries ago:

> If you plant for a year, plant a seed.
> If for ten years, plant a tree.
> If for a hundred years, teach the people.
> When you sow a seed once, you will
> reap a single harvest.
> When you plant a tree, you will reap
> ten harvests.
> When you teach the people, you will
> reap a hundred harvests.

Kuan-Tzu's poem turned out to be not so much a celebration of America's economic progress as an eerie prophecy about things to come, for in 1985 a far different interpretation of America's situation was published. The President's Commission on Industrial Competitiveness had just spent a year studying new realities of global competition faced by American industry both at home and abroad with the charge of making recommendations on improving the United States' ability to compete. The Commission reported that since 1960 ". . . our productivity growth has been dismal—outstripped by almost all our trading partners." Between 1960 and 1983, productivity (measured as real gross domestic product per employed person) in the United States changed an average annual amount of 1.2 percent, compared with 5.3 percent for Korea, 3.7 percent for France, 3.4 percent for Germany, and 5.9 percent for Japan. Thus, Japan's productivity growth has been almost five times our own.

Just as Servan-Schreiber's facts and figures had painted a portrait of economic prowess and superiority, the Commission's findings suggested the reverse, a nation unable to prevent economic opportunity from slipping through its fingers. America's standard of living, the Commission noted, has grown more slowly than that of our trading partners, with real hourly compensation virtually stagnant since 1973 (declining in 1979). Real rates of return on manufacturing assets have declined over the past twenty years. And the trade balance, positive until 1971, is now in deficit at record levels.

The erosion of the trade balance in the high-technology sector of the economy has been particularly alarming. The decline in older "smokestack" industries has been viewed by many as a natural migration of these industries to countries with lower labor costs and access to raw materials. Many argue that the U.S. economic base is rightfully shifting to high-technology industries where a competitive advantage exists. Yet, since 1980 when high-technology industries produced a $27 million trade surplus, the balance has declined each year. According to the President's Commission:

> In industry after industry, U.S. firms are losing world market share. Even in high technology—often referred to as "sunrise" industries—the United States has lost world market share in 7 out of 10 sectors. (1985, p. 13)

Education, a factor cited by Servan-Schreiber as a major contributor to America's "success," has not fared well either. The Commission reported that university revenues do not cover the rising costs of research. One-tenth of the nation's engineering faculty positions are vacant because faculty salaries do not compete with private industry. Elsewhere, studies have shown that the supply of new business doctorates has declined 20 percent since 1975. Twenty percent of authorized positions for doctorally qualified faculty across the board remain unfilled. In the information technologies area, the problem is even more acute: 26 percent of tenure-track faculty positions are vacant.

As expected, the Commission identified a number of areas that needed to be addressed if the United States were to be effective in international competition. The factors they cited were a familiar litany:

- create incentives for private research and development
- remove regulatory constraints which inhibit innovation and commercialization
- provide educational support for science and engineering
- change tax and regulatory policies which distort capital flows
- assist the workforce to adapt to change
- improve trade policies

Few of the recommendations were new and many have a lineage that dates back twenty years or so.

The Commission's report contained one disturbing omission: technology management. "Technological advance depends on virtuosity in management," Servan-Schreiber had said, yet any direct consideration of management's role was almost wholly absent from this and other treatises on the problem.

In 1986, one year after the President's Commission on Industrial Competitiveness published its report, a survey was conducted by researchers at Fordham University's Graduate School of Business. The Fordham survey was sent to 1500 CEOs of the United States' largest firms. The executives who responded had quite a different story to tell when asked why the United States has lost competitive advantage in the international marketplace, especially in technology-related areas. The executives were offered a menu of factors that might have caused loss of national competitiveness. Most of the factors were similar to those identified by the President's Commission (e.g., insufficient research and development, distortive tax and regulatory policies, etc.). Another factor was added: "Executives have not properly managed technology." Table 1-1 presents the results from the survey. According to these results, the nation's top executives believe that the one factor overlooked by the President's Commission and its predecessors ranks as the leading problem. One-third ranked improper technology management as being the single most important factor. And, three-fourths (73 percent) said it ranked as one of the three most important out of a list of nine factors; in other words, in the top one-third.

The growing significance of managing technology effectively echoes the growing importance of information technology in compe-

Table 1-1. Why the United States Has Lost Competitive Advantage in the International Marketplace

Reason	Most Important	Top Three in Importance*
1. Executives have not properly managed technology.	32%	73%
2. Tax and regulatory policies have distorted capital flows.	22	60
3. Research and development have not been sufficient.	20	63
4. Labor unions have resisted technological progress.	20	52
5. Trade policies have been unfair.	18	45
6. Elementary and secondary education systems have done an inferior job of educating.	16	38
7. Government controls have thwarted progress.	7	46
8. Employers lack incentives to invest in retraining employees.	7	37
9. Colleges and universities have done an inferior job of educating technology managers.	7	37

**Respondents could rank more than one factor as first, second, etc.*

tition. The President's Commission left little doubt about its significance:

> Technology propels our economy forward. Without doubt, it has been our strongest competitive advantage. Innovation has created whole new industries and the renewal of existing ones. State-of-the-art products have commanded premium prices in world markets, and technological advances have spurred productivity gains. Thus, America owes much of its standard of living to U.S. preeminance in technology. (1985, p. 18)

Executives who responded to the Fordham survey agreed. Most said that unless their companies moved ahead to adopt technological innovations soon, business would be lost to competitors. Eighty-four percent claimed that they would make major investments in new technology in the near future and that these investments would be made regardless of plans by competitors to do the same. The executives also believe that information technology will have a fundamental impact on the business. Three-fourths said it would change the product offered the customer. Respondents generally agreed that technology would change relationships with customers and suppliers, would significantly improve managerial and professional productivity, and would help maintain their position as a cost leader in the industry. Using information technologies to find market niche was also important.

So pervasive was the role of information technologies, that more than one-half (55 percent) agreed that it would completely change the structure of their industry. It is hardly surprising then, that over 90 percent agreed that senior managers need to become familiar with specific technology concepts to manage the business; a task which, they said, should not be left to subordinates. Over half also agreed that boards of directors should be thoroughly familiar with new and emerging technologies likely to affect the company's competitive position. To these executives, leaders of many of the nation's largest firms, technology management has come of age.

Aside from the viewpoints of the executives who responded to the Fordham survey, what are the symptoms that information technologies are being inadequately managed? There are three (see Figure 1-1). First, contrary to popular belief that managers are being deluged with innovations much faster than they can handle them, the rate of technological absorption into U.S. businesses is too slow. Second, when and where technological implementation is under-

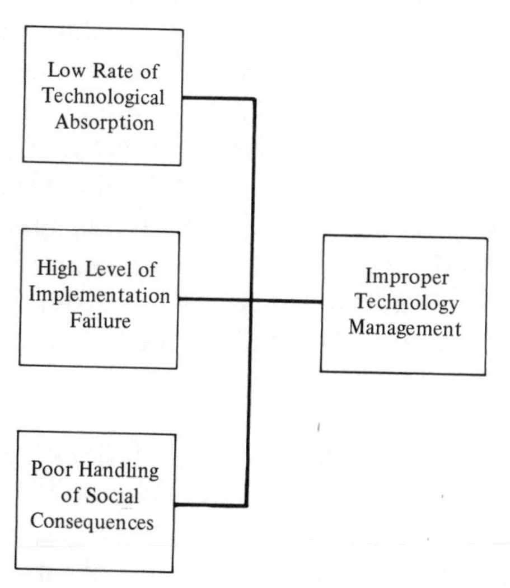

Figure 1-1. Symptoms of Improper Technology Management.

taken, the track record is disappointing; there have been too many failures. And third, social consequences—the impact on the organization and individual worker—have frequently been shunted aside to the detriment of the company's ability to extract benefits from the technology.

Technological Absorption

Undeniably, for more than four decades since World War II, we have been in the throes of a technological revolution dominated by the computer and characterized by rapid change. One observer described this era as driven by the "hurricane force" of technology. Managers keenly feel the pressure from these changes. When asked in the Fordham survey to comment on their greatest frustrations concerning technology management, a large number of respondents complained about the rapid rate of change. Respondents found it difficult to predict the course of technology and plan accordingly. They felt that determining long-range priorities therefore becomes exceedingly difficult. Rapid change introduces confusion about when to buy into a new technology and when to discard an old one.

Thus, many managers have come to believe that they are scrambling to keep up with technological innovations in vain. No matter how hard a manager may try, absorbing new technology as quickly as it develops seems impossible. As we shall see, the course of technological development is quite predictable. But the general belief by management that technology's course of development is difficult to predict is a chief contributing factor in the failure to readily absorb new technology. American businesses do not struggle to reach the technological frontier so much as dally behind it. Evidence is difficult to quantify, but many observers estimate that even large companies are running from two to four years behind leading-edge developments in office automation. Despite varying predictions that by 1990 between 65 and 90 percent of all white-collar workers would have a computer workstation, as late as 1985 only about 3 percent actually did. And adoption of robots in the United States has lagged considerably behind Japan and Sweden measured on a per capita basis.

We have come to understand that when nations lag behind in technology, the consequences can be severe. Overall such a nation would be closed out of competition in certain industries, it would become generally less competitive and less able to hold its own in international bargaining. Loss of competitiveness ultimately translates into loss of jobs and lower standard of living. Much the same effect applies to the individual company level as well.

Generally speaking, the ability to rapidly absorb new technology is desirable. Given the exigencies of today's evolving global economy, American managers want to be aggressive in adopting technological innovation. But many of today's companies, were designed for a day and age when rapid technological change was not as pervasive. This means that organizations must be redesigned to encourage and accommodate innovation and change. Part of this process involves identifying and removing obstacles that impede healthy technological absorption in American companies; to speed the rate of progress with information technologies when and where appropriate. Our problem is not that rapid technological changes are outdistancing our ability to keep pace, but that managers are failing to cultivate these abilities.

Performance Failures

More discouraging is the fact that when efforts to implement new technology are undertaken, the failure rate is distressingly high. Often these failures are blamed on poor technical design or administrative error. And while these assessments may be accurate in and of themselves, they frequently ignore the large issue of management support and involvement that provides the framework for technical and administrative activities.

Examples abound of technology systems that are prone to failure or that fall disappointingly short of desired results. For every situation where there is public awareness of the failure, numerous others remain cloaked in secrecy. Consider the following:

- A major Wall Street investment house hired a top manager who promised a new system that would integrate front and back office procedures. Millions of dollars and a couple of years later, as it was becoming obvious that the manager could not deliver, he suddenly left to go to a rival firm, taking many of his key employees. His original employer was left with a mess, highly dependent on an unworkable automated processing and data communications system, wondering what to do next.
- The New Jersey Division of Motor Vehicles engaged Price Wa-

terhouse, a major consulting and audit firm, to design a new vehicle and driver registration information system. The new system apparently performed so poorly that the division was unable to issue notices for vehicle inspection for a period of time. The result was that for weeks the vehicle inspection stations, notorious for long lines of automobiles awaiting inspections, sat virtually empty. In 1985, the Division sued Price Waterhouse for $6.5 million in damages.

- A major bank was forced to publish paid advertisements publicly apologizing to its customers for delays and inconveniences created by a changeover to a new branchwide computer system. The new system—intended to respond to customer needs faster and more efficiently—began its life by alienating those it was intended to serve.

Other examples abound. The official blame for most of these failures is laid at the doorstep of technologists, who are responsible for the technical aspects. But while many problems are in fact the result of technical failure, the overriding problem is that management permitted conditions to exist which allowed technical ineptitude and error to affect the business. In some cases, they actually encouraged technical deficiency by setting up unworkable circumstances for the technologists.

Social Consequences

Given the magnitude and scope of social and workplace changes implied by technological change, it is hard to imagine a human resource issue of greater consequence for modern managers. Technology has had and will continue to have an enormous impact on what work must be done, how it will be accomplished, and whether people or machines will perform the tasks. On a larger scale, information technologies are changing the quality of life. These changes are inexorable. Yet as a society we do not have a clear vision of the benefits to be derived from technology, or even whether the benefits are worth the perceived social costs. On one hand, people believe that technology produces all things good; on the other that it produces only evil. David F. Noble summed up this phenomenon as follows:

> Technology has been feared as a threat to pastoral innocence and extolled as the core of republican virtue. It has been assailed as the harbinger of unemployment and social disintegration, and touted

> as the creator of jobs and the key to prosperity and social stability. It has been condemned as the cause of environmental decay, yet heralded as the only guarantor of ecological integrity. It had been denounced as the hand maiden of exploitation and tyranny, and championed as the vehicle of emancipation and greater democracy. It has been targeted as the silent cause of war, and acclaimed as the preserver of peace. It had been reviled as the modern enslaver of mankind, and worshipped as the supreme expression of mankind's freedom and power. (1984, p. xii)

Social ambivalence has a greater impact on the ability to manage technologies than is commonly realized. Ambivalence robs commitment and produces inaction. Thus it is not uncommon to find managers who shy away from an aggressive pro-technology position unless forced into that stance by circumstances. Although the hypothesis remains untested, managers may take less than an aggressive posture simply because of personal feelings that technology has had negative social consequences such as displacing jobs and deskilling work. In some cases, outright antagonism toward technology prevails. During interviews with numerous managers at all levels I have observed occasional situations where a manager will find or invent a reason why information technology will not contribute to his or her business unit's productivity or goals. Often the reasons for this viewpoint relate to the perceived impact of the proposed technology on fellow employees.

Technology has negative as well as positive impacts on people. We accept technology because the benefits generally far outweigh the costs. But when the cost is measured in terms of human well-being—job security, job satisfaction, personal fulfillment—a need exists consciously and forthrightly to manage that impact. Essentially, the management task is to ameliorate the human cost and to do this, that cost must be well understood and a management plan created to avoid or control it. Currently, managers deal with the human costs by ignoring them for the most part. The result is that the workforce and society—especially those who are older, less educated, members of minorities—fear and resist technological progress. Managers for their part respond by slowing technology's progress in an attempt simply to avoid the issue or by forcing implementation of technology without adequate consideration of the social and workplace consequences. And experience is clear on this latter point: attempts to implement computer technology where key organizational changes are not simultaneously occurring are doomed from the beginning.

Taken together, these three problems—slow technological absorption, performance failure, and avoidance of social consequences—characterize the current status of American technology management practices. Certainly they interact and affect one another. Poorly implemented systems adversely affect employees, reinforcing ambivalent feelings about the consequences to the workforce. Success in implementing technology is likely to whet the appetite for a greater rate of absorption, just as bad experiences will surely dampen management's enthusiasm.

This latter point is of keen interest. Companies that have a long history of using technology successfully are much more likely to be accepting of technological change and innovation. Other companies, which may still have a critical stake in harnessing technology's energy to compete successfully, may face obstacles simply because past experiences within the organization have not been successful. A previous experience may have produced lackluster benefits or failed due to technical incompetence. Whatever the reason, the company must overcome its past experience in order to move ahead.

All three problems derive from one fact: throughout the so-called technology revolution, managers, especially senior managers, have remained without an acceptable body of knowledge on technology management. Here I make a distinction between technical issues characteristic of computer sciences and information systems, and the broader issues of managing information technology as an organizational resource of consequence. The literature that is available is scant and largely unacceptable.

Factors Contributing to The Crisis

How did we get to where we are today in technology management? Our discussion leads us to consider five major factors that have contributed to the situation (Figure 1–2):

1. *Absence of Motive.* Much of our current perspective concerning management of information technologies is governed by our historical experiences following World War II when commercial computer-based systems came into use. Our success was so complete, adequately documented by Servan-Schreiber, that we saw little need to alter existing management practices. Yet the assumption on which these practices were based were changing.

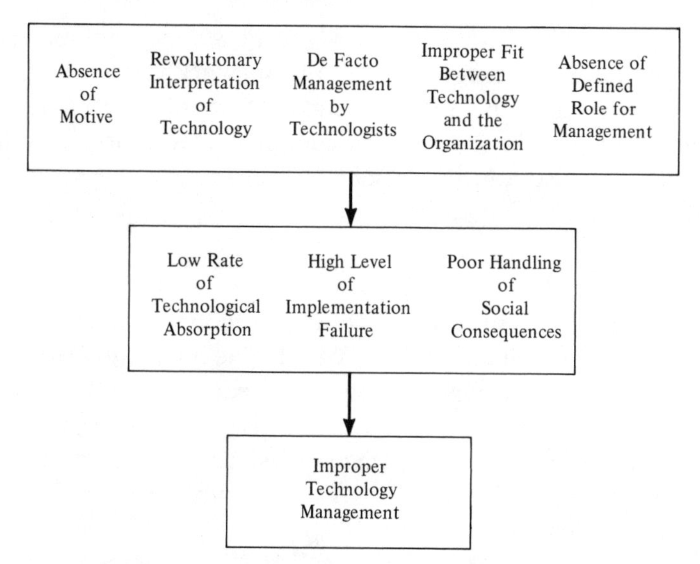

Figure 1–2. Factors Contributing to Improper Technology Management.

2. *"Revolutionary" Interpretations of Technology.* We have persisted in allowing ourselves to be blinded by technical detail. By focusing on end products of the technological innovation process rather than its fundamental aspects—those that are constant—we have concluded that new technologies create the need for new ways to manage. While management practices may have changed, the *principles* have not. Yet, fundamental principles have frequently been abandoned.
3. *De Facto Management by Technologists.* The belief, especially in earlier days, that technology should be controlled by technologists and did not require management involvement created a legacy that we live with today. That legacy, born of attention to technical detail and characterized by de facto corporate policy formulation from a technical perspective, is still with us. What is at stake can be summed up by the question: What is the rightful role of senior management in information technology?
4. *Improper Fit Between Technology and the Organization.* We are also attempting to reverse many of our initial experiences with computer-based technologies which were characterized by deep frustrations. These frustrations were a logical consequence of early characteristics of computers and how they were used. As we move away from this early era, we are left with many of the frustrations.

5. *Absence of Defined Role for Management.* More than what we should not be doing, we need a framework for thinking about what we should be doing to manage technology effectively. As we shall see, this framework cannot be developed in isolation but must also address broader issues related to how we compete.

Absence of Motive

Many aspects of current technology were languishing in various stages of research prior to World War II. The intense efforts to marshal all resources to bring technology to bear on the all-out defense effort spurred development of several major technologies, television and computers being among them. Modern computers were first used by the military during the war to do ballistics calculations and solve other problems.

World War II was the ultimate challenge to a young powerful nation which after the ordeal emerged as the undisputed military and economic power. Following the surprise attack on Pearl Harbor, the nation, still in the throes of the devastating Great Depression, was able to dramatically convert its industrial base into wartime production. After the devastation of the Pacific fleet, building the armed forces into combat readiness was a monumental challenge, all successfully met and accomplished. The glow of success continued following the conclusion of the war. Instead of returning to a depression economy, as some had predicted, the United States converted its wartime industrial base to consumer and capital goods production. What generally followed was a period of impressive growth and prosperity.

American managers certainly must have felt tremendous pride for their role in the conversion of the economy to wartime and back again. Indeed, success was so complete that few, if any, could have argued that changes in the approach to technology management were needed, even had this need been recognized. There was little apparent incentive to tamper with the nation's powerful industrial system.

But while Americans began to fall into complacency because of their success, the seeds of decline were unwittingly sown. Japan's economy had been virtually annihilated, and many of the European nations, both Allied and Axis, were in not much better condition as the result of wartime devastation. As these nations began to rebuild, many with infusions of American scientific and technological expertise, they did so using current and more productive methods. These foreign nations were forced to develop management skills appropriate to this technology. They were forced to do this because they had begun following World War II with little or nothing, and they had to challenge every assumption and reexamine each management method along the way. They built a new world for themselves not only based on newer, more advanced and productive technologies, but also created a world where rapid technological change was the accepted environment.

For Americans, change and technology were largely treated as separate questions. Change was regarded as the disruptive by-product of catastrophic social events: the Depression, World War II, and conversion to peacetime economy. Technology, on the other hand, was simply a tool. If the need for or benefit of the tool was overwhelmingly obvious, if existing methods were clearly obsolete, then new technology was implemented. Given this philosophy, American business did not cry out for new technology to be applied in existing industries, an observation supported in part by the flagging rate of civilian research and development expenditures as a percent of gross national product compared to major trading partners. The economy was good, businesses were profitable, and the American industrial base had just gone through a refurbishment during the war.

Good management of the economy during the decades since the war would have forced America's businesses to begin aggressively incorporating new technology into their operations immediately. The need for rejuvenation, even in the face of success, is a familiar strategy as illustrated in the following:

> . . . even successful high-technology firms sometimes feel the need to be rejuvenated periodically to avoid technological stagnation. In the mid-1960's for example, IBM appeared to have little reason for major change. The company had a near monopoly in the computer mainframe industry. Its two principal products—the 1401 at the low end of the market and the 7090 at the high end—accounted for over two-thirds of its industry's sales. Yet, in one move the company obsoleted both product lines (as well as others) and redefined the rules of competition for decades to come by simultaneously introducing six compatible models of the "System 360" based on proprietary hybrid integrated circuits.
>
> During the same period, GM, whose dominance of the U.S. auto industry approached IBM's dominance of the computer mainframe

> industry, stoutly resisted such a rejuvenation. Instead it became more and more centralized and inflexible. Yet, GM was also once a high-technology company. (Maidique and Hayes, 1984, p. 21)

Following World War II, had the United States realized that the situation was really a question of living off momentum generated in the past and that strong foreign competition was becoming a factor, management undoubtedly would have pursued a different course of action. The question at hand, of course, is why American managers did not recognize the need for change. Part of the reason can be attributed to the regulatory orientation of the U.S. government to business. Through legislation such as the federal antitrust laws, government had become preoccupied with the form and procedure of economic operation. This perspective was born in the early decades of this century with the need to regulate the misdeeds of unscrupulous businesses. It was reinforced by the economic upheavals of the 1920s and 1930s, and persisted throughout and after World War II. Governments of nations that industrialized later, Japan being the premier example, were forced to play a somewhat different role; they were more active in promoting industrialization and competition.

Japan moved ahead rapidly not because it had superior technical skills nor because of cultural mandates, though both may have influenced the outcome. Instead it succeeded because it had a motive to apply technology to its needs and this motive created a need to contemporize its management methods. The Japanese were forced to master technology management because of the mandate to rebuild the economy. Americans were simply not faced with such a deterministic challenge. Part of the evidence lies in the fact that aspects of current Japanese management methods have been adapted from American practices. For example, W. Edward Deming, who began his career in statistical quality control, went to Japan following World War II as part of the reconstruction effort. With his assistance, Japanese industry developed management practices that emphasized quality and productivity. Not surprisingly, the growing competitive strength of Japan has led many American managers and others to become ardent observers of these management practices.

Revolutionary Interpretation of Technology

The second factor contributing to improper technology management lies in the interpretation Americans give to technology. Since the beginning, we have repeatedly reminded ourselves that we are in the midst of a "computer revolution" or "information revolution." By definition, revolution means sudden, radical, or complete change. We tell ourselves constantly that we are buffeted by the winds of such change. The consequences are far-reaching, as the following excerpts from management literature remind us:

> As business managers we are revolutionizing the procedures of our factories and offices with automation. . . . (Hurni, 1955, p. 49)

> It has only been during the past ten years or so that the issue of mechanizing work had really emerged from the area of narrow concern . . . to become a top-management problem embracing virtually all sections of business. (McNeil, 1948, p. 492)

> Executives in several companies expressed the opinion that the greatest advantage of electronic data processing is its potential for providing new information for management. Such information will result from (a) ability to handle vast quantities of data hitherto too expensive to process, and (b) ability to carry through more complex mathematical techniques of business management than can now be practically applied. (Laubach and Thompson, 1955, p. 125)

Statements such as these, coupled with current advertisements for computers, office automation products, telecommunications, and related services evoke images of a business world in turmoil, continuously confronted with the need for radical action to reap the opportunities offered by each "new wave" of innovations while at the same time trying to steer clear of the pitfalls. But consider for a moment the fact that these statements were not written during the 1980s to describe the impact of integrated office systems, computerized manufacturing, decision support systems, or any of the other concepts in the current vernacular. The first and third were published in 1955 and the second in 1948—forty years ago—when fewer than one hundred computers were in use in the commercial sector.

At the level of historical significance, we can hardly deny that the computer has had and is continuing to have a "revolutionary" impact on business and society. After all, historians measure events such as industrial revolutions in decades if not centuries. But when we consider the nature of time constraints faced by managers attempting practical planning, we are forced to conclude that to deal with a period of thirty to forty years in revolutionary terms is meaningless. This is precisely the interpretation that many American managers bring to information technologies. In doing so, they rob themselves of the opportunity to define clearly the relevant management issues that have emerged over this period, and thereby to discriminate between important concerns where technology is involved.

Perhaps revolution is an appropriate term for what has been and is happening with computer technology over the past thirty-five years, in the sense of a second industrial revolution viewed from an historical perspective. But when one attempts to adapt this revolutionary perspective to management practices at the individual firm level, it turns out not to be a very useful interpretation. Still, many managers have consciously or unconsciously extrapolated this revolutionary view from the level of historical or social significance to ongoing operations within the company. This has a devastating effect on management's attitudes towards technology in two ways. First, subscribing to a revolutionary view implies that past management experience is obsolesced along with out-of-date business practices. A manager with substantial experience is led by this interpretation to mistrust his or her past experience. Subliminally, the message is that the experience is no longer relevant. According to this view, a seasoned manager is not much better equipped to manage technology than a newer, inexperienced manager.

The second effect is even more insidious. If one accepts revolutionary change as a given, and if one believes that change makes past management practices obsolete and that new methods are needed, then one is likely to conclude that attempting to manage technological change is a fruitless process. No sooner are innovative methods put into effect than they too are obsolete. In many organizations which, like most, are controlled by older managers who have not had exposure to computers, much of the failure to move ahead with technology can be attributed to the effects of this revolutionary perspective. Middle managers who are waiting in the wings to move into top jobs as these senior executives retire often express frustrations that derive from the firm's loss of ability to compete because of outdated production methods and senior managers who refuse to adopt new technology. Essentially, they are sitting out their tenure in office awaiting retirement, leaving the task to their successors.

Ironically, removing the revolutionary interpretation of technology as a series of cataclysmic events without rhyme or reason allows a much more dependable vision to emerge. In reality, information technology has a predictable and influenceable development or migration route. It is evolutionary in nature, and because the evolutionary process can be predicted—with minor exceptions—appropriate actions can be taken to smooth the course of organizational efforts to absorb new technological developments. So well do the Japanese understand this principle, for example, that they have undertaken to develop systems that will convert the nation to digital telecommunications by the mid 1990s. In addition, they are developing a satellite communications system that will communicate to numerous time zones simultaneously, and are attempting to create true machine intelligence. In some cases, these projects are expected to cost hundreds of billions of dollars and will require one to three decades to execute.

American managers are coming to understand that even at the individual company level such vision is needed for effective technology management. But significant barriers exist within the United States at both the government and corporate levels which prevent the cultivation of long-range technology vision. We will explore these barriers in due course. In the meantime, we persist in clinging to a revolutionary interpretation of technological development because it masks inadequacies in our methods of doing business.

De Facto Management by Technologists

The third factor contributing to improper management has been the lack of senior management involvement with technology, especially during its early days and, to some lesser extent, even today. Originally, most managers believed that computer technology would have only an isolated and inconsequential impact on the business. And because computers were a technical device, senior managers believed that the primary concern was technical. It was only during the 1960s and 1970s when computer costs began to rocket upwards and consume larger and larger shares of the corporate budget that many managers felt compelled to become involved. There were exceptions, of course. Computer projects that were likely to make a large impact on company performance were scrutinized by senior managers more closely, but involvement in the sense that technology was seen as a primary role and responsibility of management did not exist. The view was that the field belonged to computer specialists or technologists.

Initially, an element of truth resided in this viewpoint. Computers were limited and their applications were confined to the status of "glorified calculator." The question was where to put the computer, organizationally speaking. The answer usually was with the accountants, on the rather lame reasoning that accountants were responsible for number crunching, something computers seemed to do extraordinarily well. It is hardly surprising that many of the first applications of computer technology in corporations and government were accounting-related (e.g., payroll, payables, receivables, general ledger). Gradually, the computer's reputation for taming voluminous paper flows began to spread and nonaccounting applications (e.g., inventory management, insurance claims processing) became more commonplace.

Early control of computing by accountants created a misguided legacy for all those who would follow. Attempts to manage computing generally focused on the issue of control rather than on innovation and experimentation. The accepted wisdom became that computer technology should be justified by rigorous cost-benefit analysis. And these analyses, in turn, assumed that technological development is a precise and efficient process, each new development building linearly on previous developments. In fact, though technology's course is predictable over the long run, it is also characterized by many short-term detours and deviations, some of which are successful, some not.

One cannot help but wonder what the effect would have been if accountants had not been as instrumental in the early stages of commercial computing; if instead researchers or marketers had taken the lead. But such was not the case. To compound the problem, as computing began to outgrow its first organizational home—accounting—the responsibility was assumed by an even more narrowly focused group, the technologists. Throughout this initial period, the basic computer skill of interest was the creation of instructions to guide the machines's actions—programming. Programmers were often banded together informally and worked as a small, closely knit, adventuresome team. Among older managers today, there are a number who served on these teams. And during this period the primary management need was simply direct supervision. As computing moved into the 1960s, its growing use and complexity began to present increasingly difficult coordination problems. As more and more work procedures were automated and the organization became increasingly dependent on the computer, scheduling of the machines became more political. Formal analysis and design techniques began to be applied. The small cabal of programmers gave way to a more structured, burgeoning computer organization with numerous levels: systems analysts, designers, data base administrators, data librarians, and more. An era of the "computer priesthood" was in its ascendancy.

During this period, specialists remained firmly in control of the computer. Nontechnical managers came to regard computers as something better left to people who understood what they were and how they could be used. Computer specialists themselves believed that promotion from within was the desirable path to follow. Computer programmers who excelled at their task were promoted to become systems analysts. Systems analysts became project managers and project managers became directors of management information systems (MIS). This neat system of upward mobility had a serious flaw: The skills developed as a programmer were not necessarily useful as a systems analyst, project manager, or MIS director. It was classic Peter Principle.

What skills were lacking? The skills that serve the computer programmer well may be only marginally useful to a systems analyst. At that level, analytical skills—including the ability to develop alternative approaches in problem solving—become essential. Because the analyst must deal with people—not just the cold, electronic logic of the machine—emphasis on interpersonal communications skills comes into play. As a project manager, the ability to coordinate people and resources in a complex sequence of events becomes valuable. At the managerial level, an understanding of underlying technology is still important, but less so. Other, nontechnical skills are clearly needed. Finally, at the peak of this pyramid, technical skills are the least important. General management skills—leadership, resource acquisition and allocation, interpersonal communications—are the primary needs. Yet individuals with the preparation and mindset of computer programmers found themselves in positions demanding general management experience.

Vestiges of this promotion scheme remain. In recent years, much attention has been focused on the role of "chief information officer." Many assume this position will be occupied by a technologist. Yet mounting evidence suggests that while an understanding of information technologies is important, it is much less valuable than the ability to communicate, consulting skills, strategic thinking, and the like.

It is not surprising then that computer applications were—and still are—often out of sync with organizational needs. As we shall discuss later, there are certain inherent features of computer technology that dominated this period and that created frustrations for its users. But, it is also clear that some of the computer failures that occurred were due to the inability of computer specialists to communicate and to manage.

Just the difference in perspectives between technologists and users guaranteed a certain amount of communications failures. Users are generally "married" to the organization and they have a stake in seeing the organization achieve its desired results. They seek a workable solution, and one that is "quick and dirty" is usually acceptable—so long as it achieves the objective. For this reason, users are likely to be somewhat change resistant. This does not mean that they reject change out-of-hand but rather that they eschew change solely for change's sake. In other words, "if it's not broken, don't fix it." Because the user sees his or her career as depending on the organization's success, he or she is likely to be oriented more towards the organization and its culture. Although these people may employ short cuts and know people who can pull strings, they respect the hierarchy and tend to work within it.

Technologists, on the other hand, tend to be different. Because the demand for computer-related skills has typically been so great, job hopping often nets these specialists lucrative advances in terms of salary, benefits, and promotion. Thus, specialists tend to be more profession- than organization-oriented. Their interest is using the organization to develop personal skills which will be marketable in the churning computer personnel market. Elegant solutions—those that develop skills with newer, advanced hardware and software tools—are prized over "quick and dirty" (albeit practical) solutions.

But another aspect of specialists works against making them part of mainstream organization thinking, and that is their analytical systems orientation. To some extent, systems specialists are change agents who prune the organization or dissect it, adding technology and perhaps altering workforce needs. Boxes on the organization chart are meant to be moved about, broken, or combined as the design dictates. The subtleties of organizational protocol are bound to be chaffed as a result.

From a communications viewpoint, technologists tend to work in isolation. And by leaving management of the technology to these specialists, the stage was set for some disappointments. Organizations found themselves saddled with computer systems that were estranged from their users' needs and that required large infusions of unplanned-for effort to maintain them. Systems drowned their presumed benefactors in oceans of data, much of which was aggregated incorrectly for the intended use. Worse, data was simply irrelevant.

Over the years, workers spent endless hours trying to figure out what to do with the data, filing away interminable computer listings that would be rarely consulted, or pretending to use them.

With time, technologists became more sophisticated at planning and developing computer systems. Through the use of formal systems methodologies and similar mechanisms attempts were made to choose courses of action with technology carefully. But for all practical purposes, technologists were left to rely on their own resources to determine uses for the technology. Since their perspective was strictly from the technology viewpoint, applications were usually chosen because they could be done well with the technology at hand, not because they needed doing particularly. Yet, their choices about technology influenced the course of organizational events and, in that sense, they became de facto policy makers with far more power than anyone had consciously intended. Often they steered technology on a path that was not really relevant to the mission of the firm. But without direct, proactive management involvement, badly needed policy guidance was largely absent, and, as we shall later see, this led to serious suboptimal decisions about technology. Peter G. W. Keen sums the point up as follows:

> Policy is not the same as planning. Every large firm has already made, and continues to make, massive commitments to office technology, dealer order-entry systems, computer-aided design and manufacturing, electronic banking, and the like. None of these are casually undertaken. A variety of planning methodologies, as well as people, are used to carry out plans. But policy should drive plans. Policy is the set of explicit, top level mandates and directives that provide criteria for planning and selecting specific applications of [information technologies]. Policy should be simple, but not simplistic. It is management's statement about the ways things will be done. Most firms have not been resolving [information technology] questions at the top. *The management process has been marked by delegation to a technical cadre that, in the past, was largely isolated from the mainstream of the organization and had a focused, but narrow, style of thinking that left out behavioral issues* [emphasis added]. (1980, p. 35)

Improper Fit Between Technology and the Organization

To some extent, the nature of computer technology itself during early stages of commercial application contributed to improper management. Despite the emphasis in current technologies on features such as personal workstations and distributed processing, mainframe computing still dominates. And, for the foreseeable future, large computers will continue to play a major role either because a specific need for their use exists or because a company has chosen mainframes over alternative technologies for strategic reasons. We will return to these issues later. But only within the past few years has there been any real choice between large centralized data processing facilities and anything else. Until then mainframes were really the only show in town.

Readers who, like myself, remember those days also remember the considerable frustrations that accompanied them. Because data could not be retrieved directly, as is currently possible, cumbersome manipulations were needed to retrieve specific elements. The absence of computer terminals which now permit instantaneous access meant that data entry and retrieval and other functions had to be executed by writing out instructions, converting them into machine code through keypunch or similar means, and then processing by batch. And, since few software packages were available in earlier days, the process of coding programs for these machines was complicated and time consuming.

These features of early, mainframe technology created real frustrations and inefficiencies for corporations. Divisional managers with functional, regional, or product responsibilities were often forced to rely upon computing as a key strategic resource to accomplish a specific program initiative. Yet because the computer was a *centralized* resource managed by technologists, priorities were set from a *centralized technology* perspective. In the early 1970s, I remember the arrival of a young, ambitious credit manager at Exxon's Retail Credit Card Center where I was working as a systems analyst. His initiatives invariably demanded data analysis possible only through the computer since the data itself resided with the machine. But his "hot" priorities cooled to a lukewarm status as they were shuffled into the computer manager's queue of waiting projects. Early computers meant a loss of control over priorities for many nontechnical managers.

Other frustrations were present. Once a request for computer support was approved, lead time for systems development stretched for months, maybe years. Once a system was developed and installed, processing turnaround time often took days, weeks, months. The inflexibility of this early automation environment would not accommodate corporate needs. The priorities were too often the computer's rather than those of the business. It was as if a straitjacket were placed around the organization forcing it into a shape that was neither comfortable nor useful. Yet companies endured the inconveniences and frustrations for two reasons. First, even the automation of simple tasks such as payroll or insurance claims processing had a way of making seas of clerks vanish almost overnight. Large cost-savings were available and management was willing to endure much to secure these benefits. Second, computers opened the door to numerous opportunities that were otherwise unavailable. The credit card industry as we know it today would not exist without computers.

Originally, these early computers were applied in a piecemeal fashion to various unrelated tasks. Payroll and receivables systems were installed with little thought given to the need for eventual integration with other processes. But systems designers quickly began to catch on to the value of not only automating individual applications but also integrating them. Thus, we soon began to see accounting systems that shared the same underlying data coding structure for payroll, receivables, and general ledger.

From a business perspective, most managers would have seen that integration has its limits. After a point, no value remains in stringing together different applications; integrating for integration's sake. Had senior managers been in the saddle at this point, the idea would not likely have survived much longer. But, with the "high priests of tech" in control, the systems cult escalated into a religion: the totally integrated systems approach, as it came to be known. The concept underlying totally integrated systems was appealing in its deceptive simplicity. Every information flow in the organization, went its reasoning, has value throughout the organization if it can only be captured and packaged for consumption. The task therefore is to garner relevant information, summarize and filter it, and pass it upwards to management for planning, decision making, and control purposes. Thus, computer systems became management information systems (MIS) and EDP (electronic data processing) became the MIS department.

According to the MIS philosophy, totally integrated systems would be built by first capturing detailed data from day-to-day operational activities. Once this portion was in place, reports would be created by filtering out data that presumably would be of little interest to higher echelons of management. The remainder was summarized and condensed. Theoretically, this system of management reports would satisfy the needs of top executives as well as first-line supervisors. From their executive aeries, top managers could peruse the computer printouts, omnisciently controlling all organizational activities. Later, the scenario was updated somewhat when direct access to data through computer terminals became feasible. A systems specialist would thrill at the thought of the company's president consulting the report he or she had designed. The art of management revealed!

But few of the organizations that made significant investments in totally integrated systems succeeded. Many got to first base, installing the portion of the system that pertained to the lowest level of reports, those needed to control day-to-day operations at the bottom of the organizational pyramid. Once operational, these were to be followed by implementation of reports for middle management which relied upon the first set for data. Yet, companies faltered at this point. The systems rarely met the needs of their users and the expense of maintaining the systems became burdensome. In many cases, the concept of the system was simply abandoned, often after years of frustrated development efforts and attempts to make them work.

Why did they fail? Part of the reason lies in the magnitude of the task. By trying to design systems to meet everyone's needs, none were met. Covering so much territory meant that all aspects of the systems were superficial. As the business of the organization naturally evolved, basic weaknesses in the design quickly became apparent. A more fundamental problem existed. Appealing as the totally integrated concept was (and still is for some), it was based on a fatal misconception: that the primary information need of upper-level managers is summarized and filtered operational data on a routine basis. While data of this type may meet some management information needs, the proportion is generally small. In fact, we find that information needs vary dramatically up and down the organization ladder. For example, whereas operational managers' data needs are largely internal, information used by senior managers often comes from outside the corporation. Whereas operational managers are likely to need currently generated, detailed data with a high degree of accuracy, senior managers are likely to need general data relating to the future. Generally speaking, summarized and filtered data from below fulfilled only a fraction of management's information needs. Yet technologists continued to foist upon managers reams of data of small relevance.

Thus, the technology—because of its centralized nature—created a situation in which it was easy to subvert normal corporate processes and rationalize the results. Technologists, through no real fault of their own, were left in an ivory tower to reflect upon their successes and ruminate on possible future successes. And in the absence of meaningful senior management involvement, these technologists had the freedom to implement their vision, limited though it may have been.

Absence of a Defined Role for Management

The myth that rapid technological change is creating a need to revolutionize management has left a vacuum in our thinking about what should be done to manage information technologies. Granted, many companies have become quite sophisticated in planning, and in recent years a growing awareness of technology's intimate connection with business strategy has begun to emerge. Concerns like Manufacturers Hanover Trust Company in New York and others have created internal teams to consider these strategic information systems issues. But no clear idea exists about what constitutes the body of knowledge about *information technology management.* In order to unlock the potential of computer-based technologies, a paradigm that provides a positive role model, a beginning for management thought is needed. That, of course, is the purpose of this book.

We shall start by rejecting the premise that revolutionary change in management and its practices is needed. Indeed, the premise advanced here is that our willingness to accept a revolutionary viewpoint has led us astray in the past. This does not preclude change in management behavior, and specific recommendations will be made throughout the book about areas where that may be appropriate. But we will embark on our exploration by accepting the *traditional* roles and responsibilities of management as a framework for looking at information technologies. And, we shall depart by noting that among its many responsibilities, management must provide for continuity of the enterprise.

It is becoming clear that to unlock the promises of technology—competitive manufacturing, preservation of knowledge, faster organizational response, greater productivity—senior American managers must change their perspective about managing technology. Most importantly, executives must feel free to discard the notion that managing technological change involves revolutionary or innovative management practices. Nothing could be further from the truth. Indeed, accepting a revolutionary viewpoint has, in the past, misled us. A more useful approach is to ask ourselves a few simple questions. After at least three decades of rapid technological change, a time period spanning most of a senior executive's career, have we learned anything about successfully managing technology? Have experiences been similar throughout? Can we distill what we know into useful principles? Will these principles offer guidance for future generations of managers who are also faced with a pace of technological change that may dwarf what we are currently experiencing?

The answer to all these questions is "yes." We will reexamine management's role in technology and refine the definition of management responsibilities. We will abandon the revolutionary interpretation and the insistent cries for new management practices except where they are needed. We will attempt to outline major principles which should guide management's governance of technology, especially at senior levels. If the book is successful in its mission, then the reader should come away with an enduring framework for analyzing technology related decisions in business. Naturally, this framework will evolve with time as a function of personal maturity and society's expanding knowledge of technology itself. But for many the framework will be a starting place. Even seasoned executives, including those who have spent their careers immersed in technology, will find opportunities to refresh their perspective.

An Integrated Perspective

Over the past few years, managers in many of America's leading corporations have begun to think more deeply about how information technologies should be managed. Recognition is growing that these technologies are key to our success in international competition and that senior executives cannot leave to their successors the task of moving forward in this area. Recognition is also growing that American managers need to do a better job of managing technology if for no other reason than computer resources are becoming an ever greater share of the average company's asset base. Even five years ago, for example, computer-based technologies represented only a small fraction of most corporations' capital budgets. Today that figure is much higher.

These companies have been attempting to develop for themselves a complete information technology management perspective. As yet, such a perspective does not exist, for there is no commonly accepted body of knowledge as in accounting, finance, marketing, production, and management. Business schools, for example, teach a widely disparate set of practices ranging from how to use a computer as a quantitative tool to developing computer systems. Only recently have issues relevant to management such as those covered here begun to creep into the curriculum.

Companies have been left pretty much to their own devices in trying to piece together a conceptual framework for managing technology. Most have begun by consulting leading technology gurus, borrowing bits and pieces of their thinking and freely adapting elements of this thinking to their individual needs and corporate culture. In a sense, this book simply extends that process by borrowing and blending elements from four major sources. First, some useful management knowledge had been presented by practitioners and academics. This knowledge is usually focused on specific issues and often addresses particular technological developments. Yet if one looks at the sweep of thought presented in this format, two striking trends emerge. One is that throughout the history of commercial computing, certain themes have remained constant even when the form of the technology changes. The second is that we actually know a great deal about managing technology; we have reached a critical mass in our experience.

The second source of information about managing technology comes from managers themselves. Some data was collected through the Fordham survey, previously described. This survey will be referred to repeatedly throughout the book. But, a far more valuable source was numerous interviews with various company executives who were themselves deeply engrossed in concerns over the same technology management issues. Although I have gathered data from numerous companies, interviews from managers at The Travelers Companies (Hartford, Connecticut), Manufacturers Hanover Trust Company (New York), IBM (Armonk, New York), and Westinghouse (Pittsburgh) have been most influential.

A third source has been my own personal and professional experience. Many of the lessons I learned as a systems analyst and designer at Exxon's Retail Credit Card Center in the early 1970s have been surprisingly enduring over the years, an observation that I believe supports a basic premise of the book. That premise is that sound technology management comes from common-sense application of familiar management roles and responsibilities, not through embracing faddish approaches to management. The tasks of technology managers are as traditional as a button-down Brooks Brothers shirt.

Finally, the fourth source of information comes from *non-computer* technology management concepts. Engineers have, for years, studied technology management. Yet, as computer-based technologies moved away from the management control of technical specialists into the arena of general (nontechnical) management, what seems to have been overlooked is that some technology management methodologies already exist.

Plan for the Book

What all this means is that the time is ripe to assay our experiences with information technologies and attempt to bring them into a single, unified paradigm. That, in a nutshell, is the purpose of this book. The paradigm proposed is the *Integrated Technology Management Framework* and it provides the structure for the discussions that follow throughout the book. Chapter 2 is devoted to the presentation of the Framework.

The Framework is subdivided into three distinct phases. Phase I is assessment, and concerns management's understanding of the technology and its evolution, a process referred to herein as the "migration of technology." Assessment also concerns how technology will affect the industry and the company. Position-taking is Phase II; it involves a sequence of activities which lead management to decide on the level and type of commitment to technology. Position-taking is based on the important assumption that information technologies should be managed as an *investment,* not an expenditure. Phase III is policy formulation, which addresses the need to manage change in the organization, the workforce, and the larger business environment. We have long recognized, for example, that the ability or inability to change organizational culture is the chief constraining factor in adopting new technology. As one New York executive succinctly put it: "The technical stuff is easy." Yet we have done little in practice to carry through on this knowledge. Effectively managing technology concerns organizational change *as much as or more than* technology per se.

The specific objectives of Chapter 2 include:

- presenting the Integrated Technology Management Framework, the goal of which is to bring the company's technostructure into equilibrium with competitive needs

- explaining the Framework's three major phases (assessment, position-taking, and policy formulation)
- highlighting critical issues embodied in the Framework, including technology assessment, strategic uses of technology, economics of technology, creation of appropriate technostructure, impact on organization and workforce, industrial policy, and related issues
- focusing on decisions where striking the correct balance is key to effectively managing technology

Phase I. Assessment

Chapters 3 and 4 both relate to the assessment phase of the Integrated Technology Management Framework. Since the inception of commercial computing, top executives have been constantly exhorted to become involved in technology-related decisions. Many have; far too many have not. Arthur R. Taylor, Dean of Fordham University's Graduate School of Business Administration, former president of CBS, Inc., and an active board member, is succinct in his assessment. "These guys are in their late fifties, only a few years away from retirement, facing million dollar retirement funds. What they don't want to do is rock the boat by introducing a bunch of new technology."

The growing significance of information technology for accomplishing corporate business objectives has, of course, left little doubt that senior managers must be involved and must understand technology. In Peter G. W. Keen's words:

> Just as [information technology] managers need to build a more sophisticated understanding of the business, so must executives and decision makers gain a more sophisticated sense of the technology—and the actual and potential organizational consequences and opportunities it produces—if they are truly to be people who manage, execute, and decide. *Sophisticated* is not the same as *detailed.* The real issue is how little, not how much, managers must understand about [information technologies]. It has become fashionable to talk about the need for "computer literacy" in schools and in business. It is really too late for that. Managers need *computer fluency.* Literacy usually adds up to little more than confidence building through a crash course on personal computers. It narrows the issues to the visible aspects of the technology: hardware, software, and individual uses. Computer fluency relates to understanding choices and consequences. (Keen, 1985, p. 36)

Michael Millican, Business Editor of Associated Press, commented that "In times of turbulence people want and need to know more." Mr. Millican was referring to the role of media in helping business people monitor and assess the complex, rapidly changing business environment. However, the same sentiment applies specifically to technology. The issue thus becomes what management actions are needed to deal with, to assess, rapid changes in technology.

Chapter 3, The Migration of Technology, takes up the issue of how predictable the course of technology change is. The thesis is simple: managers often flounder in their understanding because they pay attention to specific technological changes rather than the big picture. A second reason is that many managers still retain a perspective about technology that was formed during the early days of mainframe computing.

Chapter 3 provides something of a retrospective of developments in information technologies to date. More specifically, it emphasizes how early computer technology was a poor fit with organizational processes and how it has now "migrated" to a form which can be molded to fit natural organizational processes.

Specifically, the objectives of Chapter 3 are to:

- encourage the view that the migration of information technologies is a predictable process that allows for long-range planning
- propose that managers should not attempt to achieve an ideal state of technology as often recommended by experts
- propose that the goal of management is to develop an "appropriate" technology infrastructure
- explain the disadvantages, including the poor fit between early central processing and the organization
- outline the technological developments that led to distributed processing and demonstrate how this newer environment should be used by management in assessing technology
- explain how information technologies can restructure the business

Chapter 4, Assessing Strategic Opportunities, focuses on the issue of how managers should cope with rapid new developments in technology. The thesis of the chapter is that companies should and can do more to research new technologies and understand their implications, and transfer that knowledge to units within the organization where it can be beneficial. The greater the impact information technology has on an industry, the more important it becomes in place to assess technological change and advise management.

A specific technology assessment methodology is proposed that recognizes interaction between business goals and technology; business goals are as much affected by technology as they affect adoption of technology. The methodology emphasizes forecasting the impact on industry, and identifying and analyzing potential changes in industry structure. Only then can various alternative courses of action to the firm be considered.

Specific objectives of Chapter 4 are to:

- emphasize the need for managers to seize the initiative in assessing new and emerging technologies, especially when the technology is likely to change industry structure
- critique current technology assessment practices
- provide examples of how leading technology companies create centers of expertise in specific technologies and transfer that expertise into operating units
- explain a general technology assessment methodology

Together, Chapters 3 and 4 set the stage for the next phase in the Integrated Technology Management Framework: position-taking.

Phase II. Position-Taking

Chapter 5, The New Economics, delves into the economic issues in technology management. It confronts what may be the most serious obstacle to effectively managing technology: our lack of understanding of how newer, communications-based technologies change the economics of scale and scope. In Chapter 5, I argue that communications-based technologies differ in economic performance from earlier information systems. Essentially, the technostructure must reach a critical mass of automation before the real benefits can be obtained.

Managers who invest in office automation but do not understand this economic behavior are often disappointed with the results or put off by the inability to quantify and forecast an acceptable level of benefit. In turn, these experiences lead to frustration with the technology and cause managers to make insufficient investments.

Specific objectives of Chapter 5 are to:

- review traditional methods of measuring and evaluating economic performance of information technologies
- analyze the changes that were introduced by communications-based technology which invalidated assumptions underlying traditional performance measures
- consider the consequences of continuing to use traditional measures which are inappropriate for newer, communications-based technologies
- suggest alternative approaches for evaluating performance
- identify barriers to improving technology's performance over the long term.

Chapter 6, The Technology Investment, examines general courses of action available to management. Armed with an understanding of how technology can affect industry structure, management must now consider how it will use technology to compete, if at all. Specifically, Chapter 6 begins with a discussion of the goal of information technology management which is to build a business technostructure consistent with the company's competitive needs. After beginning with fundamental assumptions about the basis of competition for the firm, the chapter explores different positions that can be taken in response.

The major issue that emerges is that to compete effectively, management must build up a general technological capability. Thus, the major decision to be made is the division of resources between two competing needs: immediate applications and general long-term capability. Chapter 6 concludes by discussing the need to have a developed architecture or plan for the business technostructure.

Specific issues dealt with in Chapter 6 are to:

- demonstrate how technology should be matched to the general competitive strategy for the company
- differentiate four major positions that can be taken in response to the company's competitive needs
- explain how each of the four responses mandates different decisions about technology
- focus on the need to develop a business technostructure to provide an appropriate level of capability and flexibility
- consider the trade-off that must be made in terms of using

investment dollars to satisfy immediate application development needs versus improving the business technostructure.

Phase III. Policy Formulation

Chapters 7, 8, and 9 all relate to policy formulation, the third phase of the Integrated Technology Management Framework. Chapter 7, Technology and the Organization, addresses the process of introducing organizational change as a means of facilitating technological innovation. The chapter begins by arguing that hierarchical and bureaucratic organizations will not disappear as some observers suggest. Instead, they will be modified to accommodate new demands placed upon them by the competitive environment.

Chapter 7 also examines how foreign competition is beginning to alter the basis of competition to focus more and more on customer satisfaction through improved quality and responsiveness to marketplace demand. This in turn influences management's decisions about how to employ technology. The overall policy goal for management is to synchronize the organization's culture with its competitive goals. Then, once appropriate organizational issues are resolved, technology systems must also be synchronized.

Specific issues dealt with in Chapter 7 are to:

- demonstrate that organizations will retain their familiar hierarchical and bureaucratic form
- explain forces being shaped by foreign competitors which are changing the competitive missions of many American companies
- emphasize the need to match technology systems to appropriate organizational form
- consider methods of accomplishing organizational change as a means of facilitating technological innovation.

Chapter 8, Technology and the Workforce, deals with issues of what information technologies mean to the individual workers as a means of helping management shape appropriate policies. For a variety of reasons, people have ambivalent feelings about technology, but the potential for strong bonds between the two are more powerful than commonly thought. Managers who fail to understand this basic principle are likely to overlook important advantages from the technology.

Specifically, issues taken up in Chapter 8 are to:

- explore the complexities and potential influence that technology systems can have over employee behavior
- consider the evidence regarding both long-term and more immediate impacts on employment levels
- examine how information technologies might restructure the workforce, especially with regard to the growth of a "contingent" work population
- consider how management and other work might be transformed by information technologies and what need this creates for policy formulation

Chapter 9, The Environmental Context, completes the trilogy of chapters related to policy formulation (Phase III of the Integrated Technology Management Framework). This chapter relates to factors outside the organization—and therefore beyond the direct control of management—but which must be taken into account. Three categories of factors are considered: (1) accounting and financial measurements, (2) industrial policy, and (3) business education and research.

Specifically, issues explored in Chapter 9 are to:

- explain limitations of existing accounting and financial controls
- propose actions that can be taken to remedy or partially remedy shortcomings of these measures
- explore the need for a coordinated national policy to improve the corporate environment for technological innovation
- consider other public policy changes that would improve the likelihood that American managers would make effective technology-related decisions
- examine the role of business education and research for enhancing our understanding of technology and related management practices.

Chapter 10 concludes by synthesizing the principles that underlie the Integrated Technology Management Framework. These principles, founded on traditional management roles and responsibilities, emphasize a *balance* between the three phases of the Framework: assessment, position-taking, and policy formulation.

2

The Integrated Technology Management Framework

Begin with Quality

The paradigm that will guide our discussion throughout the book is the Integrated Technology Management Framework presented in Figure 2–1. This Framework is an attempt to satisfy the needs for a single perspective at the senior level regarding the management of information technologies in the contemporary competitive environment. The last phrase is key since, as we shall see, the lackluster results companies have often experienced with information technologies relate more to the objectives being pursued than to success of implementation. The Framework differs from previous thinking about information technology management in three ways. First, it rejects many of the revolutionary interpretations currently being advocated by other writers. The Framework was developed in part by examining the traditional roles and responsibilities of management and relating them to information technology issues. One of the results of this process is that few concerns for technical questions emerged. Instead, the Framework focuses on issues that are characteristic of the broader set of concerns that routinely occupy management's attention. The Framework represents a return to traditional

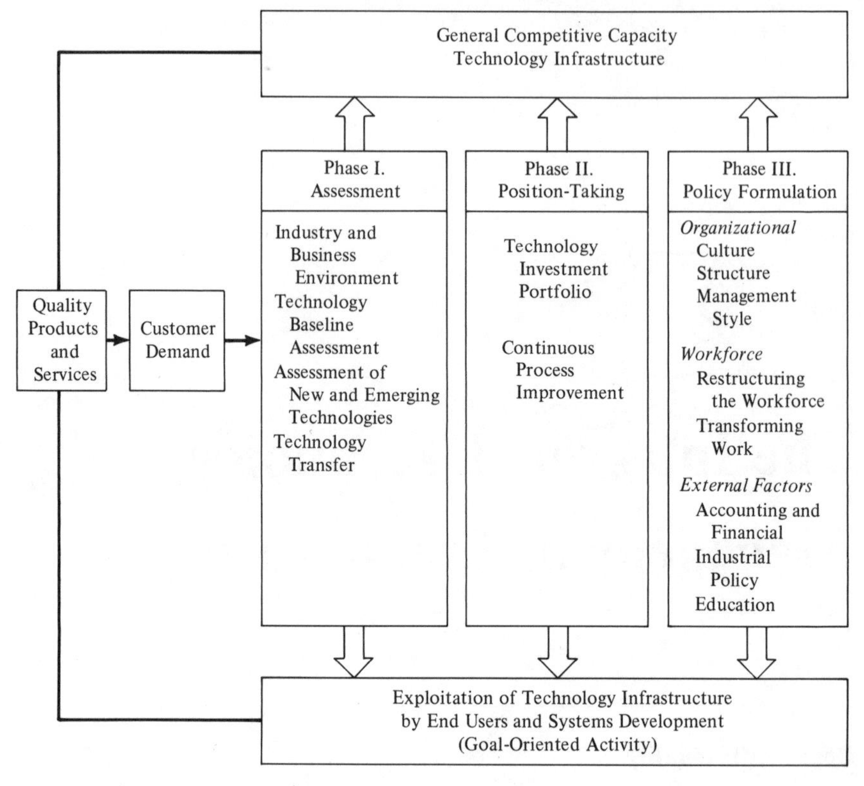

Figure 2–1. The Integrated Technology Management Framework.

thinking in other ways as well. For example, based on an understanding of the purposes of human organization, it flatly rejects notions advanced by some observers that computer technology will put an end to hierarchical structures and bureaucratic organizations.

The "integrated" part of the Integrated Technology Management Framework refers primarily to the fusion of technological issues with traditional management thinking. In the past, we have tended to view technology as something separate and apart from other organizational concerns. As noted in Chapter 1, computer management was thought to be the preserve of technologists, not business managers. Therefore, what people previously defined as management issues have in reality been *technical* questions. These technical concerns have filtered up to senior management, sometimes appropriately and sometimes not. In many cases, senior managers felt that they had little or no frame of reference for making these decisions. These questions seemed to have little relevance to their own concerns, yet they seemed representative of the preoccupations of subordinates responsible for computing resources. Senior managers have lacked a framework of their own for technology-related decisions, which gives rise to the "framework" part of the Integrated Technology Management Framework.

The second reason why the Framework differs from previous approaches is that it explicitly incorporates mechanisms to deal with management's understanding of new and emerging technologies, organizational and workforce issues, and factors external to the firm. We have given these factors lip service from the start, yet little of substance has been done to address them. Our experience tells us, over and over, that these factors are at least as important as the technical and investment decisions. No matter how well the technology itself is executed, the effect is to throw seed on barren ground if the innovative process has not taken root within the organization and the workforce. The same is true where the firm faces constraints that emanate from the external environment which effectively nullify technology-related decisions. In this sense, the Framework is "integrated" because it combines the management of organizational issues related to technological innovation and implementation with the technical issues. We have never had a management paradigm that dealt with these factors in a forthright manner.

The third difference relates to the goal of information technology management. American managers have always regarded the computer primarily as a tool to cut costs and increase productivity. In recent years another, supposedly more enlightened, view has emerged which sees information technology in a broader role. According to this view, information technologies can contribute to any activity within the firm's value chain; they can, for example, create new and differentiated products and services. Dun & Bradstreet, for example, uses computerized data bases to tailor corporate credit reports to meet the particular specification of a client. The technology can, of course, also be used to fill a specific market niche. Information technologies can also be used to help a company be a low-cost producer. In other words, information technology can be used as a strategic weapon to capture competitive advantage in a number of ways.

The problem with this expanded view is that it really provides no guidance about how information technologies should be put to use; it does little more than recite technology's potential contributions which are, of course, as countless as the grains of sand on the beach.

Part of the shortcoming in the "technology-as-strategic-weapon" philosophy can be attributed to American managers' imperfect understanding of how to compete in general, a topic which will be explored in more depth in due course. In brief, competition is often looked at as a process of identifying goals and objectives, designing programs to achieve them, and then executing them. In the past, computer technology has more or less been managed consistently with this perspective. Little in the nature of the physical technology contradicted this view. Computer systems could be built as discrete, objective-oriented entitites.

Yet, with the advent of communications-based technologies such as office automation and computer-integrated manufacturing, the nature of the technology changed. An objective-oriented perspective thereafter made little sense as a complete interpretation of how to compete with technology. Although goals and objectives must always be a central part of competitive or strategic thinking, opportunistic capability is as important. Opportunistic capability, essentially a short- and medium-term issue, means being positioned to respond rapidly to changes in market conditions which open opportunities for exploitation. Companies like The Travelers Companies and McGraw-Hill, Inc. provide ideal examples which will be explained more fully later. The Integrated Technology Management Framework is based on this opportunistic approach. It says, in effect, that senior management's task is to maintain a technological base or infrastructure which provides the company with the appropriate striking capability.

The technology infrastructure is much like an aircraft carrier from which aircraft launch assaults on selected land-based targets. Without the vessel, the aircraft cannot get within striking distance and are therefore useless. In the past, with information technologies, senior managers have focused on the aircraft and tactical missions. The Framework says that senior managers should provide the aircraft carrier—the technology infrastructure—and let the pilots—the business unit managers—worry about accomplishing specific objectives.

Underlying the shift away from an objective-driven view of competition to an opportunistic view is the knowledge that technological evolution is not an efficient process. The nature of technological innovation and development is inherently risky, sometimes resulting in dead ends and rapid obsolescence of even recent developments. Desktop computers are now available with the power of mainframes of the early 1980s. This means that technology is more in tune with the opportunistic approach, for it seeks to create general competitive capability. To some extent, the objective-driven view assumes that technological development is an efficient, controllable process. This view has been relentlessly reinforced by the accountant's perspective which quickly savages any application of technology that fails to contribute immediately and measurably to the bottom line. This interpretation is clearly unnatural when one considers the process of technological innovation. The Integrated Technology Management Framework is more realistic because it is explicitly based on the opportunistic perspective and recognizes that the most lucrative benefits may be in the innovative and creative aspects of the technology.

The argument being advanced is that current methods of measuring profitability in American business are fundamentally incompatible with the economic realities of the technological development process. These measures, as will be shown in good time, distort and subvert the decision-making process. Here we encounter a real dilemma. Any approach to managing technology that uses accepted measures of profitability as its goal is not consistent with economic realities. Yet, any approach that is out of sync with current profitability requirements is not likely to curry much favor among American managers no matter how accurately it reflects the technological development process. The net effect is that if we force technology-related decisions to conform to current measures of profitability, suboptimal outcomes are the result. In other words, information technologies do not perform up to their full potential under these circumstances and, as noted in Chapter 1, this malaise is common.

Until recently, American business could live with this malaise without serious side effects. But that is no longer the case. Two factors have joined forces to change the situation. One is that the underlying economic behavior of information technologies has shifted with the growing importance of communications. Previously, benefits could be derived with the piecemeal application of computers. But now, evidence suggests, a critical mass of automation within the organization must be reached before many of the expected benefits can be successfully derived. Reaching critical mass is key. This is likely to involve substantial investment in technology over an extended period and it is likely to involve having a conscious design for the technology infrastructure from the outset. IBM's Lexington, Kentucky manufacturing facility is a case in point. Approximately five to six years were required to fully convert it from the traditional mode of manufacturing to computer-integrated manufacturing.

General Motors' Saturn plant being built in Tennessee was started from scratch. Its time horizon for obtaining the benefits of interest will undoubtedly be much longer. Similar logic applies to the application of technology to the office environment.

The second factor is growing competition from abroad, especially from countries such as Japan. Companies from these nations have been redefining the basis of international markets with their particular approach to competition which differs from the American approach. Americans look for a fairly immediate return on investment, of course, and thus decisions are number-driven. Decisions about quality of product are judged on bottom line contribution. The Japanese, on the other hand, are not generally concerned with the numbers. They focus on satisfying customer demand through quality products and services. Furthermore, they aim for *continuous* improvement of quality. Their goal is not to achieve an "acceptable" level of defect but to strive continuously for zero defects. Though they may never actually reach this goal, the constant striving towards it produces the desired result. But the reduction of defective products lessens the need to rework items, resulting in reduced production costs and increased productivity. This approach to cutting costs by emphasizing zero defects is quite different from traditional American thinking which accepts a certain level of defective production as economically optimal.

Quality, therefore, has two important effects. It creates a better product or service and therefore satisfies the customer. In the Japanese view, this creates additional demand which increases business. Quality also leads to productivity improvement. The Integrated Technology Management Framework accepts this view as its starting point. Thus, as will be explained in later chapters, the Framework offers a valuable set of tools for managing information technologies in a way that will be globally competitive and will achieve productivity gains. However, it requires a substantial reorientation away from the increasingly parochial American approach to competition. In this sense, the Integrated Technology Management Framework is not simply a method of dealing with information technologies in isolation, since doing so is counterproductive. Rather, it is a more complete management philosophy.

The intent is not to muddle the mission of this book by introducing factors extraneous to the discussion of effective technology management. But a central theme of the book is that these factors cannot be separated. However, the Framework rests on three premises that need to be specified. Each is to an extent based on the perspective used by the Japanese, and derived in part from the work of Americans such as W. Edwards Deming and others who consulted with the Japanese following World War II in an effort to help rebuild the economy. Deming was instrumental in convincing the Japanese that quality production was critical if they were to compete successfully in foreign markets.

In recent years, American managers have become increasingly interested in Japanese management practices and the "Deming approach." As might be expected, a number of consultants and academics have hopped on the bandwagon either as ardent promoters of the gospel or as creators of management methodologies that are largely derivative. In practice, most American companies that have examined these management approaches in order to better understand how to improve their own competitiveness have come away with little more than the notion that quality is important. As a result, companies now have quality awareness programs consisting for the most part of posters and other trivia. Some, such as Ford Motor Company, even require their supplier to have quality programs. So suppliers, as if complying with some noisome government regulation, dutifully trot their employees off to quality training courses which rarely motivate or otherwise modify behavior in a meaningful way. Thus nothing useful is accomplished.

What American managers miss in their assessment of the Japanese and Deming-oriented perspectives is that the issue is not whether quality products and services are being produced by American industry per se, but whether American industry has the *freedom*—indeed, the choice—to produce quality. Thus, although the Integrated Technology Management Framework heartily endorses quality as the starting point, it broadens its horizons to consider what conditions are needed to permit quality. In that sense, the Framework places its confidence in the native talents of the American managers: it says that by establishing the climate for quality, quality will follow. As we shall see, information technologies are a powerful vehicle for creating the proper conditions.

While the following premises are freely borrowed from the Japanese and the Deming schools of thought, they are adapted for their relevance to American business. And, they underlie the discussion of the Integrated Technology Management Framework. The three premises are as follows:

1. With information technologies, productivity is not an end unto itself. The goal of technology systems is to produce quality products and services, and all applications should be judged on that basis.
2. Information technology cannot be managed in isolation, but must be managed as an infrastructure. Infrastructure includes human workers, methods, management style, as well as information technology. The system must be viewed as a whole. It extends beyond legal organizational boundaries to include suppliers, customers, and others. Senior managers should focus on defining infrastructure, not implementing technology.
3. Information technologies are a central tool for continuously improving the system. When improvements are made, they are captured and reproduced by the technology system and thus counteract the effects of entropy where quality is concerned. They leverage quality improvement.

Technology Infrastructure

According to the Integrated Technology Management Framework, management's major responsibility is to build a technology infrastructure appropriate to the competitive needs of the business. However, the term "technology infrastructure" involves more than computer hardware and software, and telecommunications. It encompasses both technology and human workers in a single system of production and administration. Current thinking often considers only the technical aspects. Recently, for example, much attention has been devoted to concepts such as the "architecture" of information technologies or the "information utility." Architecture generally refers to the physical configuration of technology throughout the company and never explicitly considers human employees. It is therefore a component of overall infrastructure; thus, once senior managers deal with infrastructure issues, technologists design and implement architecture. Information utility refers to the availability of information through the architecture. The idea is that like a water or power utility, information is readily available and a system of delivery on demand is in place. Information utility as a concept is incorporated into the more encompassing infrastructure.

Current thinking also sees a relationship between technology and organizational structure. According to this viewpoint, information technology architecture influences organzational structure and structure influences architecture. But the Integrated Technology Management Framework regards structure as well as other organizational characteristics as an *explicit* part of the process of managing technology, arguing that if these components are not a part of decision making, then results are likely to be disappointing. In fact, the Framework proposes that organizational change precedes technological change. Organizational form—including structure, culture, and management style—should be matched appropriately to the business's competitive needs, then technology should be matched appropriately to organization and competitive needs.

The concept of technology infrastructure is a major departure from past thinking which has seen information technology as a tool kit of sorts to fix specific problems or to accomplish specific objectives. The main thrust of management's role in building and maintaining infrastructure is developing what the Framework calls "general competitive capacity." Rather than build information systems for a specific application, general competitive capacity positions the company to take advantage of a range of competitive opportunities. Several changes in the fundamental character of information systems have brought about the need for management to pursue this course of action, as we have already seen.

First, information technologies are undergoing a shift in their economic behavior. Until recently, isolated applications could be cost justified. A single automation project was undertaken because management expected certain benefits which were relatively easy to forecast from the start. But, as communications technology moved to stage center in manufacturing and office automation, benefits came to depend on having a critical mass of the operation automated rather than an isolated portion of it. Many benefits could not be realized until the technology infrastructure was substantially in place. This concept eluded many managers who continued to look to isolated applications of computers to produce the same benefits they had become accustomed to in the past.

Second, the problem with converting technology infrastructure is that it takes considerable time to accomplish. As already noted, IBM successfully converted a traditional production plant in Lexington, Kentucky to computer-integrated manufacturing in a process that took about six years. Similar experiences are becoming more and more commonplace as companies realize that major conversions of technology, especially when they involve substantial restructuring of traditional methods of doing business, take time. Shortening the time required is difficult because complex technology systems, often unique in design, yield numerous surprises along the development and implementation path. Experience is beginning to show us that these systems must be developed incrementally if they are to be successful and this, of course, requires time.

Taken together, what this means is that much of current information technology's benefit will be rendered only after lengthy development efforts spanning three to five years or more. Planning specific business objectives that far into the future is all but impossible. Thus, the attitude towards the strategic uses of information technology must change, focusing more and more on providing general competitive capacity for the future. As we shall see, American managers face considerable barriers in accomplishing this task. For example, accounting and financial measures encourage precisely the opposite behavior: investments that will produce short-term results rather than provide long-term capacity. Other countries do not face these constraints. Their managers not only have the freedom to invest for the long term, but are actually given encouragement and support for doing so. This is why the Integrated Technology Management Framework takes into account the need for management's overt involvement in changing factors outside the firm: accounting and financial standards, government policy, and education.

Once the appropriate technology infrastructure is in place, general competitive capacity provides a "platform" that permits short- and medium-term exploitation by business units within the firm. Without the appropriate competitive capacity, these units would be substantially less able to take advantage of opportunities, whether expected or unexpected, as they present themselves. Thus, the infrastructure provides an opportunistic capability that the firm would otherwise not have. Exploitation occurs in two major ways. First, applications can be developed through the traditional systems development process involving computer systems specialists as the need arises. This process is, of course, familiar to most managers, many of whom still regard it as the primary management responsibility where information technologies are concerned. In recent years, this approach is increasingly being confined to situations where the task is highly specialized or where the complexity and magnitude exceed the capacity of the nontechnical business unit. The second mode of exploitation is through the end user. End users are nontechnical professionals, administrators, managers, and others who can personally employ computers to satisfy their needs without involving significant technical support.

Technology infrastructure plays another role. It is the principal vehicle by which quality in products and services is maintained. Continuous improvement of the general competitive capacity of the infrastructure yields two important benefits. First, the quality improvement is incorporated forevermore into the system. It will continue to produce quality long after the effort to create it is completed. In essence, quality is institutionalized. And second, technology systems *can be* a powerful tool for setting quality standards and motivating workers to respond to them. As we shall see, the bonds between workers and technology are strong and it is up to management to see that these bonds develop constructively.

Three Phases

At the simplest level, the Integrated Technology Management Framework deals with these issues in three distinct phases: (1) assessment, (2) position-taking, and (3) policy formulation. These phases are not deterministic in the sense that management must work through one, then the other, and so forth. Rather each represents a particular locus of related concerns. In assessment, the Framework essentially addresses the problem of dealing with rapid technological change and what it means for the business. Assessment encompasses all activities related to gaining an understanding of information technologies. This includes developing organizational mechanisms for assessing both the company's current (baseline) capabilities with technology and new and emerging technologies. It also includes mechanisms for transferring the knowledge gained from the assessment phase to the firm as a whole. Position-taking refers to the activities related to the actual investment in information technologies. These activities include determining the potential economic performance of the chosen technology and then, based on that knowledge, deciding on a portfolio of technology investments. Position-taking attempts to explicate the trade-offs managers must make in their investment decisions and also to give direction and purpose. Finally, policy formulation concerns the management policies related to both internal and external matters which are closely associated with technological innovation and implementation. Policy formulation provides a framework for dealing with organizational change associated with technological change.

Generally speaking, any one of the three sets of concerns could as easily precede the others and each concern is as important as the others. In order to obtain the desired benefits from information technologies, management must pay just as much attention to assessment or policy formulation as it does to position-taking. Previously, these areas were often considered less important. Management's "support" for new technology somehow abrogated the need for management to understand it. The assumption was that organizational issues would take care of themselves in time. Management's emphasis was placed on the investment decision.

In a sense, the objective of the Framework is to achieve a balance or equilibrium between a number of forces. The Framework attempts to achieve a balance between the firm's future competitive needs and the technology's capacity to fulfill those needs. It attempts to achieve a balance between building future capacity and short-term exploitation of that capacity. It attempts to achieve a balance between the organization culture and the rate of technological innovation and implementation. It also seeks a balance between the need to make technology-related decisions in the long-term interest of the firm and the external factors that govern management's decision making. It attempts to achieve a balance between the rapid rate of technological change and management's willingness to invest in existing technologies. And, it attempts to balance the understanding of what technology is capable of doing with what ought to be done.

Each of the three phases is examined in more detail below.

Phase I. Assessment

The assessment phase has four identifiable components:

Assessment of industry and business environment

Self-assessment to determine "baseline" of technology

Assessment of new and emerging technologies

Technology transfer

The first component, assessing the industry and business environment, relates to the normal process of collecting general intelligence in preparation for planning. Thus, a broad range of factors—economic, market, nature of competition, industry structure, financial—are included. The importance of assessing general industry and business factors is that technological developments of themselves are rarely sufficient to create major change in the approach to competition. Real forward movement with technology in an industry is generated when other elements of the business environment in that industry are also undergoing reformation. Airlines were the first major example. They had reached a size and complexity in the 1960s that mandated improvements in reservations handling if growth were not to be crimped. During this period interactive computing was just coming into its own. Interactive computing allowed a reservation for an airline seat to be recorded immediately through a computer terminal, thus removing any confusion at any point in time about whether the seat had been sold or not. Thus, changes in both the nature and evolution of the industry and the technology were occurring and proved to be symbiotic.

The second example is provided by the financial services industry. Inexpensive large-volume transaction processing needed for growth in consumer banking and insurance, and a precondition for the rise of the credit card industry, developed simultaneously with the commercialization of computing itself. The sheer number of financial transactions needed would have been economically impossible had it not been for automated processing. Currently, the financial services industry is again crossing paths with technology. Developments include globalization of markets and deregulation of the industry within the United States. The technological development of major interest is telecommunications, since without the capacity to send data instantaneously neither globalization nor deregulation would get very far. The impact of information technologies on the financial services industry is traced in greater detail later in the book.

The second component is self-assessment to determine the "baseline" of the technology within the business. "Baseline" simply means the current status of technology within the company; the installed technology asset base. Part of what the baseline study attempts to determine is the strengths and weaknesses of the current systems. This study is really a point of departure for determining how the company should proceed.

Many senior managers do not have a firm idea of their current systems capabilities. Many do not know, for example, whether the information technology is sufficient or insufficient for the company's needs. Installed systems are often seriously underutilized, in which case managers feel it makes little sense to undertake expensive additions without first improving the performance of the existing assets. The baseline study may also yield some startling results that influence the company's course of action with technology. One example is provided by The Travelers Companies of Hartford, Connecticut. Travelers' historical roots are in the insurance business but like other financial services companies, it has been restructuring itself to offer a diversified line of financial products and services. For example, the company acquired an investment banking firm, became more aggressive in real estate investments, and made other changes.

The company had always been a big consumer of computer equipment. Historically, its primary automation need was large-volume transaction processing and its information systems had been designed to do this efficiently. But as the company and the industry moved towards integration of financial services, a different need arose: the ability quickly to produce a wide variety of financial services products and market them innovatively. This put very different demands on the company's internal computer resources. Not surprisingly, managers responsible for the system came to believe that the technology in existence at the time would have to be junked and new systems built using a technological approach more suited to the new regime. But the baseline study quickly revealed that the company had a $200 million investment in its current systems. It further revealed that the existing systems were much more serviceable than had originally been expected. Management thereafter developed a plan to adapt its existing technology infrastructure to the new competitive environment. The approach they took will be discussed more fully later in the book.

Assessing new and emerging technologies is the third component of the assessment phase. Technology assessment, as it is presented within the Integrated Technology Management Framework, is a methodology for monitoring and evaluating technological developments that is important for two major reasons. First, it provides a formal mechanism within the organization to deal with management's need to understand technology. This point cannot be emphasized enough. As Chapter 1 chronicles, management involvement with information technologies has been far too passive. Evidence suggests over and over that managers give lip service where informed action is needed. Since the beginning, we have recognized the need for a better understanding of the technology, but little has been done to ensure that this happens.

Most companies currently attempt to address this need through management training and similar programs. Yet such programs do not really resolve the need. These programs are usually conducted in a piecemeal fashion, pulling executives out of the loop, feeding them technical terminology, exhorting them to use technology as a strategic tool, giving them some "hands-on" experience, and returning them to their routine, sometimes little affected by the experience. Because technological change is an ongoing process, the need for technology-related understanding is continuous as well. A formal mechanism for technology assessment must answer to this need. Also, most companies need a more sophisticated function for digesting the torrent of data about new developments and synthesizing it into something meaningful for the company.

The second reason is that organized and active assessment of emerging technologies is likely to help companies clarify what their involvement should be further upstream in the technology development cycle. Companies that depend heavily on information technologies to improve their competitive position especially have an incentive to understand the flow and timing of new technologies, a process referred to herein as the "migration of technology." This migration route is much more predictable than is commonly believed, and Chapter 3 delves into this issue. Understanding what is occurring upstream in the technology development process and perhaps actively participating in that development could greatly contribute to improvements in technology's performance within a given time frame. As we shall see in later chapters, this approach to technology assessment sometimes makes for strange bedfellows. Market researchers at a telephone company, for example, have found technological solutions among the Department of Defense's efforts to create an artificial intelligence-driven battlefield management system.

The final component of the assessment phase is technology transfer: the seeding of important knowledge about new and emerging technologies into the organization as a whole and into specific business units. In a real sense, the transfer question opens up the larger issue of how the organization is equipped to handle innovation and change generally. Companies have tested and used a number of approaches. Training is ubiquitous but, as already noted, tends to have a spotty effect. Companies also use specific organizational mechanisms, information and technology "centers" to transfer knowledge. But overall, the need for a constant and organized transfer of technology-related information usually remains unsatisfied within most companies.

The lack of quality and commitment in these programs is responsible for more of information technology's lackluster results than most people have recognized. The reason is that transfer agents within the organization normally define their responsibility as the *cost-effective* dissemination of technological information. These agents have failed to create a real vision of technology's capabilities. Here vision refers to the ability of business managers to think through the possibilities; it is an innovative, even creative, element. This element is absent from too much of our technology-related activity. Yet innovation and experimentation are the engine of momentum in the technology development process. For lack of a better way to say it, the technology transfer function should not only disseminate accurate and reliable information, but should also set business managers' imaginations on fire. They must understand the possibilities before they can master them. And as they master them and adapt them to corporate purposes, then and only then does the time come to articulate the benefits to the business. Timing is everything.

Ultimately, the purpose of assessment is to set the stage for taking advantage of various competitive opportunities available to the firm. American managers have been notoriously weak at seeing the potential of information technologies especially as they might impact trends within industries. Historically, at least until the mid 1980s, managers remained mired in the limited and inaccurate belief that computers are largely a cost-cutting or productivity improvement tool. We are now beginning to understand that information technologies offer a much richer range of opportunities, a view we have been pressed into by developments in foreign competition. We now see that technology serves competition by creating new products and services, and by differentiating existing ones within the marketplace. We are beginning to see that industry structures are being inexorably changed by technological forces. The players change. The balance of power between customer and supplier is shifted. Barriers are erected (or demolished) for new entrants. Existing industries die; new industries are born. Sometimes the technology pervades the business so thoroughly that the two become indistinguishable. Grumman, for example, now describes aircraft as electronic components sheathed in a metallic shell. Sometimes the transformational process is so powerful that companies leave their traditional businesses and embark on new and more profitable undertakings. Singer, long known as the sewing machine company, sold that segment of its business to concentate on advanced defense technologies. Westinghouse, previously noted for consumer appliances, made a similar transition.

Phase II. Position-Taking

The second phase of the Integrated Technology Management Framework—position-taking—contains the activities traditionally emphasized in management, that is, deciding what investments will be made in information technologies. Two major decisions are really the crux of the position-taking phase. The first involves the resources that will be committed to long-term development of the technology infrastructure versus resources needed to exploit the existing infrastructure in the near term. Normally, long-term infrastructure development—or general competitive capacity—does not lend itself to traditional accounting measurement for reasons already mentioned and reviewed in Chapter 9. Basically, the infrastructure takes time to design and implement, and bring to peak performance. The process hinges not only on technical change, but on organizational change as well, and social change is often slow and may not be responsive to the application of time or money.

Changes in the nature of technology, for example the increased role of data communications in office automation and computer-integrated manufacturing, often mean that a system must be substantially converted to the new technology before benefits really begin to flow. The reasons for this phenomenon are found in the shift in economic behavior that has occurred along with the technological changes. In essence, the company must reach a critical mass. Piecemeal application of technology, which has been the approach used in the past and still reinforced by current accounting measures, is now likely to yield only suboptimal benefits. Piecemal benefits may accrue, to be sure, but they often prove to be secondary.

In some sense, critical mass benefits seem to imply an "all-or-nothing" philosophy. The system-wide infrastructure must be in place before benefits can be had. At the heart of the matter is that management may be faced with the prospect of making substantial investments over a period of years that yield marginal or "unacceptable" results in the short term, at least when judged according to present financial accounting standards.

In essence, financial performance measures in the United States contain a systematic bias against developing long-term competitive capacity with technology. With newer technology, principally that which is communications-based, foregoing infrastructure development means foregoing the most lucrative long-term benefits. We are now in the very heart of darkness where the failure of American competitiveness is concerned.

In later chapters, I will repeatedly argue that shortcomings in American management of information technologies can be attributed far less to the factors usually identified—learning curves, cultural factors, and the like—and far more to the way in which we evaluate performance. And, I will argue as forcefully as I know how, and as other people are beginning to do, that real progress can be made only by changing the basis for these evaluations. Frankly, without such changes, the nature of which will be discussed in due course, I do not see how appropriate decisions about information technologies can ever be made. Under those circumstances, our methods will become weaker and more infirm, and the evolutionary process within the economic system—only the fittest survive—will discriminate against us.

But, waiting for economic reforms, assuming that one agrees they are needed, is of little help to technology managers in the near future. What guidance is available? Essentially, the position taken by the Integrated Technology Management Framework is that senior management must structure a portfolio of technology investments that are appropriately diversified on three dimensions: risk, return, and time horizon. The portfolio approach offers two very important advantages. First, it balances investments made for the purpose of realizing future benefits with the actual realization of current benefits. Too much investment in the future may allow existing, readily available opportunities to go unheeded, although the two are not necessarily mutually exclusive. On the other hand, too much emphasis on investments for short-term exploitation fails to provide needed future capacity. Analysis of the portfolio at any given point in time yields answers about where the emphasis is being placed.

Secondly, diversification may allow management to achieve an acceptable level of return for the entire portfolio even though long-term benefits are less measurable. The reason is that it is careful to choose short-term investments that are above average in performance. Here again, management is attempting to strike a balance; this time on risk and return. There is, however, a limit on the overall performance that can be had from a portfolio because of the current counterproductive emphasis on short-term results.

The second major decision that forms the crux of the position-taking phase regards the practice of continuously improving the technological infrastructure. As already noted, continuous improvement of products and services is a major principle underlying the competitive philosophy of the Japanese, a premise accepted by the Integrated Technology Management Framework. A conscious effort to improve infrastructure is an explicit task in managing technology to the extent that the product or service is improved. In other words, improving the technology system per se may not always lead to improved products or services.

In position-taking, continuous improvement is as important as the original investment decision for two reasons. The first is that technology, like all systems, suffers entropy—a winding down of its effectiveness. Over time, the system is patched and stretched to meet evolving business conditions. As the business environment changes, as it is apt to do, the infrastructure becomes less and less relevant. The second reason relates to the effect changes to technology infrastructure have on human beings—employees, customers, suppliers, and others. Technology, as it turns out, is a powerful medium for establishing (or improving) standards of excellence and expectation. And once established, technology reproduces the standards faithfully except, as already noted, for the effects of entropy.

Investment decisions and continuous improvement serve two very different needs within position-taking. Investments are likely to be decided on a cyclical basis, usually as a part of the normal planning process. Whether related to general capacity or short-term exploitation, investment decisions are usually made in response to specific organizational goals. Continuous improvement by its very name implies an ongoing activity that embodies no other goal than improvement for improvement's sake as long as it is related to customer satisfaction. The dual nature of these activities is therefore probably best handled by separate areas of organizational responsibility.

Position-taking occurs simultaneously with assessment, and with policy formulation for that matter. While assessment systematically and continuously builds up the level of expertise within the organization as well as the understanding of various courses of action open with information technologies, position-taking mediates this knowledge by deciding how it will be used. Neither is more important, a caveat that extends to policy formulation as well.

Phase III. Policy Formulation

The third phase of the Integrated Technology Management Framework concerns the formulation of management policies oriented toward both internal and external factors which are appropriate to the technological position already taken. The development and promulgation of these policies is as key to effective technology management as the ability to assess them and make intelligent investment decisions. Policy sets the stage by preparing the organization for change. If the brief history of information technology management has taught us one lesson above all others, it is that success in managing technology depends more on organizational factors, human acceptance, and social challenge than on technical issues. Certainly technical competency comes into play eventually, but rarely will a technical decision make a major difference in the overall effectiveness of the technology infrastructure in serving business objectives. In any event, senior and even many middle managers are rarely in a position to make informed technical decisions. That task should normally be delegated to subordinates who are specialists, and who have the time and indeed the responsibility to keep current on developments within the technical field.

The guiding principle in policy formulation is that the form and content of the technology infrastructure should be matched to the form and content of the organization. If an organization is highly decentralized, then some decentralized control of information technologies is appropriate though, as we shall see, the issues are somewhat complicated. Organization form and content should, of course, be matched to corporate mission. In short, organization follows mission, technology infrastructure follows organization.

Past thinking about management has often emphasized structure and strategy. But more recent approaches have stressed the importance of balancing or bringing into equilibrium a number of components: structure, strategy, management style, systems, workforce education, and organization culture. This view is consistent with the Integrated Technology Management Framework, which essentially adds technology infrastructure as an explicit component. What the Framework says, in effect, is that forward momentum with information technologies is best created when *all* components are being advanced. Unless all variables in the organizational equation are appropriate for the current direction, then the potential for performance is compromised. Many managers mistakenly assume that the implementation of technology is separate from the process of managing the organization. Worse, some seem to believe that the implementation somehow substitutes for the process. These attitudes account for more in the way of unmet expectations from information technologies than is commonly realized.

Policy formulation has three main thrusts: organizatonal, workforce, and external factors. According to the Framework, organizations will continue to use both hierarchies and bureaucracy despite predictions by some experts that information technologies are obsolescing those forms. However, technology allows organizations to be much more responsive to market forces and entrepreneurial in their dealings. Therefore, the need exists for organization culture to reflect these characteristics as well as flexibility, adaptability, and creativity. Innovativeness will increasingly become a routine part of organizational process.

To these ends, some organizations are likely to experience a flattening of the hierarchical structure, but this is much less likely to occur than many current sources forecast. Structure, as always, will be the servant of corporate mission. Hierarchy and even the much maligned bureaucracy are useful manifestations of structure and will remain so.

In the foreseeable future, much of management's attention will need to be devoted to transforming modern organizational thinking from a static to a dynamic orientation. How is organizational change, especially *technological* change, to be incorporated into the corporate psychology? We will find that much of the key to this question lies in understanding the experimental and risk-taking nature of technological development and implementation, in effect ending the myth that these processes are efficient. Thus, the degree to which forward momentum is created will depend in part on how well managers incorporate innovativeness and creativity into the scheme.

The importance of this view can be seen in Japanese swordmaking, one of the traditional arts. Today, large metallurgy companies are known to support craftsmen who practice their ancient craft. The painstaking care which goes into the making of a sword offers inspiration concerning the dedication to excellence. And corporate sponsors also have found that the centuries-old techniques provide insights into how current mass production can be improved. Preserving an enclave of excellence and creativity ultimately has its benefits.

The second major thrust of policy formulation is the workforce. A basic premise of the Integrated Technology Management Framework is that workforce issues should be aggressively tackled well in advance of technological implementation in order to maximize improvement in information technology's performance. The problem, of course, is that despite the enormous implications for the workforce, we understand very little about the potential effects. Indeed, as it turns out, information technologies tend to be highly selective in their impact on specific industries, geographical regions, and perhaps even individual firms. As a result, we are often left with conflicting and incomplete information about their precise impact.

A major objective of the Framework is to exploit the interrelationship that develops between human beings and their technology. Under the proper circumstances, technology can be a powerful motivational and direction-setting tool. Under ill-conceived conditions, it can create resistance and even punitive actions by human workers.

PHASE

I

Assessment

3

The Migration of Technology

The Need for "Vision"

Keeping up with the rapid rate of technological change is one of the greatest frustrations faced by American managers. According to the Fordham survey, managers who commit to a new technology are often chagrined to find it obsolete within two to four years and they feel duped as a result. According to one respondent:

> Any hardware discussion today in retrospect is wrong within two years. The rate of technological change is dizzying and exhilarating at the same time.

Another problem is that few predictions about technology and its capabilities have borne fruit. Expectations often remain unmet, deadlines come and go without implementation, and performance is less than what had been promised.

Over the years, various "visions" about future technology have been advanced by experts. In the 1960s and 1970s the vision centered on the totally integrated management information system which would satisfy information needs throughout the organization. At the touch of a button, managers would be able to obtain the data they needed in the form they wanted. More recently, the growing use of communications technologies has focused the vision on office automation with its theme of better enabling knowledge-workers to perform their tasks. And in the factory, the computer-integrated manufacturing (CIM) systems concept has emerged. Experts have also begun to talk of the "computer-integrated enterprise," a company in which all activities are automated and integrated through communications technology. The result, they predict, is a more responsive and competitive business enterprise.

But these visions rarely bear any semblance to the eventual reality. As the director of strategic planning in a large New York bank observed, "In information technologies, we haven't delivered what we promised since 1960. We have made a mishmash of it." And, some truth attaches to the comment; as we discussed in Chapter 1, flaws in the totally integrated management information systems concept prevented practical implementation. Projections of penetration for office automation have been overly optimistic; since the late 1960s the billions spent on office automation equipment have not caused aggregate white-collar productivity to rise significantly, certainly not in proportion to the investment. In factory automation, progress has been slow and truly computer-integrated manufacturing systems are still rare.

Little wonder, then, that managers find it difficult to predict the future course of information technologies. From their perspective, long-range planning becomes exceedingly difficult and, some believe, even futile. Undoubtedly, these problems cause management to be less aggressive in the adoption of technology than might otherwise be the case.

Yet, when managers who successfully manage technology are asked what makes them effective, their response almost invariably includes a reference to the ability to create a "vision." What they mean, of course, is that one must have an idea of how technology will evolve, how it will manifest itself within the industry some years hence, and perhaps an idea of how the company will operate its business using technology in the future. For these managers, rapid technological change holds little intimidation value. The reason is deceptively simple: they have a sense of the grand scheme of things—a "vision"—that enables them to put individual technological developments into perspective. The vision is based on a knowledge of where information technologies have been and where they are going, a process which we will refer to as the *migration of technology*.

Dedicated word processors are one example. In the early days of office automation many managers made considerable investments in stand-alone dedicated word processors only to discover that within a brief period, word processing could be performed much less expensively using general-purpose microcomputers. Not only were microcomputers less expensive, but they were able to perform numerous other office automation tasks and were generally more flexible. Dedicated word processors were typically "incompatible"—they could not be integrated with other office automation (including word processing done on general-purpose microcomputers). Not surprisingly, many managers felt deceived by the technology and believed their decisions to be wrong in retrospect. Consequently, many have decided to sit back and wait for the technology to "shake out." Once it stabilizes, they reason, the time will be ripe to automate.

Managing technology is not unlike being the captain of a sailing vessel. In order to chart a course, the captain must know the destination. But the route is not likely to be direct. The very nature of sailing mandates that the vessel tack to and fro, making numerous course adjustments for wind, weather, sea conditions, and obstacles encountered along the way, even though each tack moves the vessel closer to the final destination. To the uninitiated passenger, this constant tacking must seem unnecessary and disorienting; each change in direction appears to contradict the previous leg of the trip. Yet the captain, who understands sailing and navigation, knows that each tack contributes to the process of reaching the final destination.

Managers unfamiliar with technology are much like the landlubbing passengers in that each new development seems to contradict those that have gone before, as in the example of dedicated word processing. As in sailing, technology often tacks to and fro by the very nature of the technical development process. In business, tremendous pressures exist to be efficient and cost-effective with information technologies. The assumption is that the technological development process itself is efficient and linear, one advance building on another. But such is not the case.

We have seen numerous developments come and go. Consider the venerable punch card, for example. At one point it was ubiquitous in computer culture; now it is virtually obsolete. Mass storage devices such as drums, and vacuum tube and transistor memories have come and gone. Each served its fleeting purpose.

It may therefore seem prudent to hold back and wait until technology "shakes out." Indeed, many managers have argued that there is little point in investing in a technology that will soon be obsolete. But remember that technology continues to move forward. Progress does not halt in its tracks waiting for the "shake-out" to occur. Even when a "shake-out" occurs, the results are temporary. New developments will soon appear on the horizon. Indeed, one change in perspective needed in technology management is for managers to spend less time sweating individual developments and focus more on trends, especially trends now on the horizon. Much of this chapter is devoted to helping the reader sort out what is trend and what is trivia. Much of Chapter 4 concerns learning how to assess future trends. Precise prediction is, of course, impossible, but those who know sea navigation or technology management understand that course adjustments are to be expected now and then.

Understanding technology's migration route is helpful for two reasons. One is that it helps predict technological trends that sometimes take decades to unfold. Dedicated word processing is really one small phase in a series of developments that are reforming how text is processed. In turn developments in text processing are part of the larger office automation evolution which in turn is helping restructure white-collar work. Like Chinese boxes, one inside the other, each single development has broader significance, and thus individual decisions regarding technology investments can be placed into context.

Second, understanding the migration of technology helps determine where the company *should be* along the route. No company is as advanced with technology as it *could* be. Only rarely can a company justify being on the cutting edge. The management issue is how far along the company should be. The investment in technology infrastructure must correspond to the need for the company to remain competitive *in the appropriate market.* And here, two implications emerge. First, the notion that a company should race ahead with technology to create a "computer-integrated enterprise" or any other showcase concept using technology makes little sense unless management can clearly articulate how this action will create and sustain an advantage over its competitors. Second, failure to maintain the appropriate pace with technology causes the company to fall behind the pack. Falling behind usually means that the technology infrastructure's capacity is insufficient to support the company's competitive needs.

The point is that the best management decision should focus on the question Are we competitive? and not Are we the technology leader? It is unimportant in itself whether American factories become totally computer-integrated manufacturing systems or that the computer-integrated enterprise, where all activities of the company are combined into a integrated, electronic whole, becomes a reality. But, it is *vital* that American managers understand what the evolving capabilities of these technologies are and how they can contribute to making their firms competitive. And as technology continues to evolve, the answers to those questions will change.

Some readers may conclude that the lesson to be gleaned from this is to run with the pack. Getting too far ahead of competitors does not pay and falling too far behind has its inherent dangers. That may be accurate to a point. For example, it is not at all clear that building computer-integrated manufacturing facilities, despite advantages over existing methods, will ever restore certain U.S. industries to a competitive status in the international arena. Thus, overreliance on a particular approach with technology, or going too far ahead of the pack, may lock the company into a technology that is not competitive. As the basis for competition in the industry shifts, the technology in place must be flexible enough to accommodate the change.

Companies do not have to get ahead of the pack to find themselves locked in. A case in point is the now defunct People Express Airlines. Started in the early 1980s, People was a revolution of sorts because of its no-frills, low-cost air travel. Not only were fares low, but the fare structure was simple. Most routes had only peak and off-peak rates.

The computerized information system that People Express developed would not permit ticket exchanges with other carriers—a practice common in the airline industry—because it could not accommodate the complex airfare structures used by the other airlines.

At first very successful, People Express soon found itself in substantial difficulties and was sold to Texas Air Corporation. Several management errors were to blame, one of which was the design of the inflexible computer system. Other airlines began to compete with People Express on the basis of cost and had the added advantage of flexible fare structures. Yet, People Express was neither ahead nor behind the industry as a whole in the application of technology.

This then, is the goal of managing technology: to bring the technology infrastructure to a level of capability and performance that will permit the company to match its competition over the long haul. No more, no less. But, what permits technology to move forward and increase its usefulness to business? Here, we must be careful to define "development"; it refers to general process or forward movement with technology in applications to business. It does not mean simply technical innovation. Indeed, developments in technology's application to business usually come about because of three key criteria: technological feasibility, business need, and economic feasibility.

Technological Feasibility

Forward movement with technology is governed in part by its technical feasibility. In many cases, a technology is technically feasible long before business applications of that technology are in demand. The Picturephone is one example. Developed by AT&T years ago, the device permitted users to see as well as converse with one another. The whole notion of seeing another party to whom one is talking some distance away has again gained currency because of efforts to integrate voice and video. The technology is beginning to be used in teleconferencing which is, of course, a specialized application when compared to casual consumer long-distance communications. This technology—in teleconferencing form—is coming into use only because all three criteria are beginning to be met.

Business Need

A technology is highly unlikely to develop if a reason for its use is absent even though it meets other criteria. In 1982, for example, Apple Computer, Inc. introduced a machine called LISA. LISA was an extremely advanced personal workstation. It had roughly four times the computing capacity of microcomputers commonly available at that time plus sophisticated screen graphics. LISA was a "desk metaphor" containing electronic versions of activities typically engaged in by managers and professionals. The machine was extremely easy to use, incorporating many features that made it a paradigm of user friendliness.

LISA was also a marketing disappointment. The machine was simply too far ahead of its time and businesses did not perceive a need. After a couple of models had been tried, the machine was still unsuccessful. Later much of the technical design was recast into the Macintosh computer. The Macintosh was more modestly priced, much more in line with other microcomputers with which it competed. The Macintosh was marketed successfully as a user-friendly, graphics machine. By the time it had arrived on the scene, managers were becoming increasingly concerned about user friendliness and graphics capabilities.

People often become very excited about new technology, especially at the personal computer level since it involves the broad populace. Amiga, an inexpensive IBM compatible microcomputer, rapidly created a following because of its graphics, color, and audio qualities. Some afficionados believed these features would cause it to scoop IBM and capture large market share in business applications. But such was not the case. At the time, the business market was concerned with mundane basics of spreadsheet analysis which are executed perfectly well without sophisticated graphics, color, and audio.

As the sophistication of business uses of computers grows, graphics (and hence color) and audio links will become commonplace. Then, need will be measured by demand. Until then, no matter how technically sophisticated or inexpensive, the technology will not fly.

Economic Feasibility

The third criterion is that the technology must be cost effective or at least perceived as cost effective. Many managers currently approve expenditures for microcomputers for use by professionals, for example, because they have "faith" that white-collar productivity will be improved. "If I pay a professional $50,000 a year plus benefits," the rationale goes, "and he tells me he needs a microcomputer costing $2,500, does it make sense to deny the request?" For these managers, perception is as good as the real thing, an issue we shall return to in Chapter 5. More often, the issue of economic feasibility is decided based on quantitative analysis. In the mid 1970s, I worked as a systems analyst for Exxon's Retail Credit Card Center when point-of-sale (POS) equipment was just beginning to take root. Exxon would routinely reevaluate this equipment for potential use in its retail outlets. Clearly a business need existed for the dealers to obtain purchase authorizations efficiently and the application was technically feasible, but the cost of the telecommunications link was then too expensive. Many things have changed in the last decade, telecommunications costs among them. Ultimately, the economics criteria were met and the point-of-sale equipment installed.

The relationship of these criteria to technology's capacity to make change is shown in Figure 3–1.

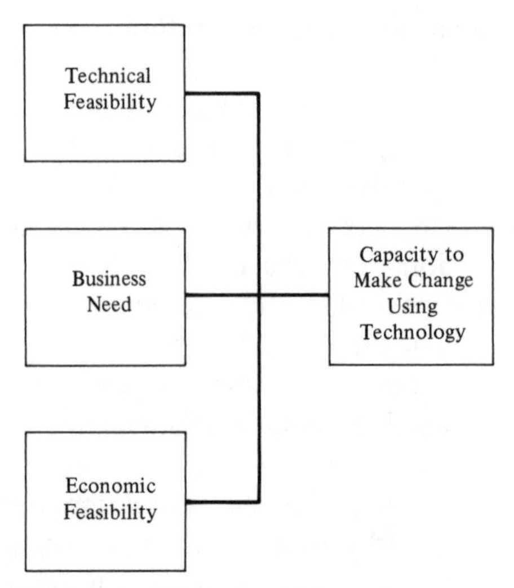

Figure 3–1. Three Criteria for Technology Movement.

Migration of Technology in Financial Services: A Case Study

Of all industries, the effects of information technology's migration can most clearly be observed in financial services. The industry has been undergoing substantial changes at all levels. Deregulation, for example, is beginning to break down traditional specializations in financial products and services, a move that is beginning to give rise to integrated financial service companies. Companies previously specializing in commercial banking (Chemical Bank in New York) or insurance (The Travelers in Hartford, Connecticut) are moving into investment banking. Deregulation has also opened the possibility of creating truly national banking to replace regional banks.

Another trend affecting the industry is the globalization of world financial markets. Investment trading is beginning to move towards twenty-four-hour-a-day operations. Once bastions of conservatism, financial institutions are placing increased emphasis on the development of new financial instruments, products, and services, and new methods of marketing. As a result disintermediation of the credit process is growing, including securitization (the switch from non-marketable to marketable debt).

These trends are intimately linked to information technologies. So strong is the link that it is difficult to tell whether technology facilitated deregulation and heightened competition or vice versa. For all practical purposes these issues are inseparable. The fact remains that the financial services industry has closely mirrored the developments in technology and has quickly adapted itself to take advantage of the available opportunities. It thus provides us with an excellent case study to observe the migration of technology.

One of the players in financial services is Manufacturers Hanover Trust Company, a New York-based bank with $75 billion in assets and $8 billion in revenue (1986). Like many other institutions, Manufacturers Hanover is evolving into a diversified financial services organization with decentralized management. In the mid 1980s it restructured itself into five strategic business sectors corresponding to its five major customer groups:

1. *Asset Based Financing.* National middle markets, large-ticket equipment financing worldwide, and real estate
2. *Investment Banking.* A wide range of investment banking services on a global basis including the origination, trading, and distribution of securities, foreign exchange, and other financial instruments, merge and acquisition financing, venture capital investments, and the provision of fee-based advisory products including investment management and specialized corporate services
3. *Retail Banking.* Deposit services, loans, leases, and credit cards to consumers and small businesses nationwide
4. *Banking and International.* Local currency and cross-border lending, trade finance activities and transaction processing services to corporate and government accounts overseas and correspondent banking worldwide
5. *Corporate Banking.* Innovative and traditional banking services, including credit, trust, securities, and transaction/information products.

Central Data Processing

In the 1960s Manufacturers Hanover, like most transactions-oriented concerns, had large mainframe computers running in batch mode in highly centralized functionally oriented operations. In the words of Edward Nyce, Executive Vice President for Manufacturers Hanover Information Technology Services, "These workhorse computers were aimed at cleaning up the back office and streamlining the processing. And to a sensational extent, they succeeded." Central data processing grew out of the earliest applications of large mainframe computers. Surrounded by easily accessible and inexpensive computer technology today, it is easy to forget that originally those machines were the only choice available. They were expensive to operate and were physically confined. To enter data or retrieve it, one had literally to go to the central computer since remote access devices like terminals were not available. The wellspring for computing, organizationally speaking, was accounting based on the assumption that since computers crunched numbers and accountants crunched numbers, there must be a fit. Predictably, the earlier applications were also accounting (and transactions) oriented: payroll, receivables, general ledger, inventory. But soon enough, others in the company began to catch on to the computer's benefits and applications spread to other functions.

Centralized processing brought with it a number of problems that resulted in suboptimal organizational performance. The following were among the disadvantages:

1. *Loss of control over business priorities.* Not only did managers fail to recognize in the early days that information technologies should be linked to business goals, computers actually caused a loss of control over priorities. Because computing resources were centrally controlled, some method of prioritizing their use was needed. Managers who headed data processing created procedures for ranking requests for applications from various functional managers throughout the company. For these functional managers this meant a loss of control. A marketing manager might need a computerized analysis of customers in order to implement a new advertising initiative, for example. But what might be a "hot" priority to the marketing person turned out to be only "lukewarm" to the technical people.

2. *Long lead time for applications development.* Another problem was related to the lead time necessary to develop new applications. Unlike contemporary technology, computers were, at that time, pretty much the exclusive domain of specialists. Analysis of needs and the design of new systems remained exclusively in the hands of programmers and systems analysts since computer technology was complex and difficult to use. But specialists were usually in short supply, thus creating a bottleneck in developing applications. Even when a specific project was given priority, many nontechnical managers were frustrated when six months, a year, two years, or more went by before the new system was operational. And, these time estimates did not include commonplace delays in implementation.

3. *Lengthy turnaround in processing.* Following the implementation, processing often took an unconscionably long time so that data often arrived days, weeks, and months after it was needed. One newly arrived manager, for example, found that budget expenditure reports were being generated *five months* after the fact, hardly in-time to provide useful feedback. Delays were sometimes caused by poor management, but more often than not lengthy turnaround time was simply a characteristic of computer processing in that era. Data had to go through a tedious and time-consuming series of steps which included preparation for input, actual processing, and output. Even with proper management, the degree of improvement possible was limited.

4. *Poor fit between the technology and the organization.* The early days of computing were, of course, dominated by large, complex mainframe computers. Organizations had one, or at most a few, to begin with and therefore little decentralization was possible. The centralized computing mandated by these large mainframes in turn often forced a centralized structure on the organization itself. Remote accounting functions attached to regional offices, for example, were closed down and transferred to the home office as record keeping was automated. This frequently had the effect of removing a vital and timely source of information from the place where it was needed most. In a real sense, the computer subverted and suboptimized many organizational processes. It prevented the organization from functioning in the best interests of the business enterprise.

This development had tremendous consequences because it made the organization less responsive in general and less responsive to its customers in particular. The misfit manifested itself in a variety of ways. For example, information was increasingly accumulated and reported not because it was wanted but because it seemed to fit an "ivory tower" conception that it *should* be useful. The result was mountains of output, mostly in the form of green and white computer printouts, which began to blanket corporate offices. Managers often felt plain stupid because of their inability to make sense of the jumble of facts and figures contained in these printouts.

As the computing function grew in size and power, the technologists in control began to flex their organizational muscles. They believed their mandate was to bring automation to the uninitiated. Because of the momentum that had developed behind the automation trend and because of the general ignorance outside their ranks, few people were in a position to stop them. Increasingly, technologists became unrealistic in their expectations about what computer technology could accomplish. They began to see the computer as a single medium through which various applications throughout the company could be *integrated* or connected one to the other. Working from a largely theoretical (and as it turned out impractical) perspective, they argued that because any process is ultimately linked with any other process within the company, a totally integrated system could be designed and built. Since all data was handled by a single centralized data processing operation, the whole idea, which later formed the basis for the management information systems (MIS) concept, hinged on first being able to automate transactions data at the lower, operational levels, and to then summarize and filter that data upwards to middle management. In turn, middle management data would be summarized and filtered upwards for senior management's consumption.

The whole notion of totally integrated information systems had tremendous intuitive appeal and many companies invested heavily. But, in practice, the approach was usually short-lived. While the lowest-level portions of the system were often successfully implemented, attempts to build middle and senior management level components ran into difficulties.

The reasons were deceptively simple. First, managers at the upper levels were not always interested in summarized data coming from below. The focus of their concerns was usually external: economic factors, competition within the industry, consumer interests, regulations. Internally oriented systems did not capture this information. The type of information in which management was interested differed in other ways as well. The accuracy and frequency required was less than for other levels and other dissimilarities existed.

The second reason for the failure was that the designers assumed that the company was static. By building systems that concentrated on the automation of *existing* procedures, they ignored the fact that companies continuously change. For example, a new marketing promotion is undertaken. A division is sold. Another company is bought and merged into existing operations. The rigid, static systems the technologists constructed were quickly obsolesced. While they were trying to repair the damage such changes did to the operational-level systems, plans to automate (and integrate) middle- and upper-level management needs languished.

Why did managers tolerate these frustrations at all? The answer is twofold. First, even for the simplest automation tasks, cost reductions were often dramatic. Accounting and bookkeeping, an obvious example, had previously been done by armies of clerks arranged in uniform rows of desks, using adding machines and tabulators to perform simple arithmetic tasks. These vanished from many companies almost overnight. Second, computers opened up numerous opportunities that simply did not exist previously. The credit card industry, as we know it, flourished; it depended directly on the ability to accomplish basic transaction processing efficiently and cheaply using computer automation.

Recently, the power and capabilities of large mainframe computer technology have become available in small, inexpensive computers—even personal computers. This phenomenon has led many managers to conclude that central data processing is being replaced by this newer technology. Such is not the case; it is supplementing central computing, not replacing it. Despite the current emphasis on office automation and other decentralized end-user concepts, central computing operations still dominate much of information technology. These mainframe computers continue to be the transactions-processing workhorses and they will continue to play a role in shepherding information. Extremely large and complex data bases husbanding massive reservoirs of information to which immediate access is often needed are growing in size and volume. These require large computers to manipulate. Communications technologies are also growing rapidly and many managers have not yet realized that significant computing power will be needed to control them. Thus, many organizations will find that the role of central computing will continue to grow despite other developments in information technology. Therefore, an understanding of the central data processing environment is still relevant to managing technology.

At a minimum, most central computing facilities will become maintenance operations for automated transactions processing which is now in technological maturity for most large companies. Here it may have a smaller share of the total responsibilities and a more specialized role. In 1984, for example, Manufacturers Hanover's operations division had 8,000 employees and a $500 million operating budget. Three years later, the core unit declined to 2,000 employees and a $150 million budget. Only 15 percent of the corporation's technology expenditure remained centralized. What led to the change?

Distributed Processing

In the 1970s, a new era of computing began to percolate into organizations: distributed processing. The essence of distributed processing was that computers had become smaller and less expensive. Where one or a few large mainframes predominated before, several smaller machines could be substituted, and where computers could not previously be cost justified, they increasingly won acceptance. Constraints that central data processing had placed on the organization began to relax somewhat even though most distributed data processing environments remained under central control. For Manufacturers Hanover, distributed processing meant that operations could be decentralized. Separate technology units for (what were then) major market segments were established—retail, wholesale, securities, and internal corporate requirements. These units—called vertically integrated data centers (VIDCs)—had dedicated computer and systems development resources. During this time a global telecommunications network (Geonet) was also built. Most large, sophisticated financial services firms were following Manufacturers Hanover's lead.

Distributed processing began as the result of three critical developments in computer technology: (1) increasing capacity at decreased cost, (2) interactive computing, and (3) telecommunications. These trends represent a mighty convergence, the impact of which will take decades to work out. *This development essentially mapped out much of the migration route for the predictable future.* Far from being events that simply happened at some point, a wave of innovation that came and went, these three developments created the potential for *fundamental* change in the structure of organization processes.

Increasing Capacity at Decreased Cost

From the beginning, computing capacity has been increasing dramatically while costs have been decreasing. Advances in this area have been so well documented that little value can be derived from elaborating on them here. Suffice to say that the basic computing power needed to do processing on mainframe computers in the mid 1970s can fit onto a single microcomputer chip now. And this rate of progress is unlikely to change in the foreseeable future. This means that computing power will eventually be regarded with about the same importance that we now attach to a sheet of paper from a writing tablet. Current costs will in retrospect seem exorbitantly high. Think, for example, of how precious papyrus scrolls were considered in ancient time. Ultimately, computing power will be extended to every nook and cranny of human existence to the point where computer capacity will be "throw-away."

The ongoing phenomenon of increasing capacity and decreasing cost is closely intertwined with the three criteria for forward movement with technology (Figure 3–1): technical feasibility, business need, and economic feasibility. As capacity increases and costs decrease, entire ranges of potential applications become feasible. This feasibility often creates the business need. Once the need is articulated, pressure builds for further technical improvements which will increase capacity and decrease costs. The cycle repeats, of course. This sequence of events has become known as "technology-push." And, nowhere can its effects be so clearly observed as in the evolution of desktop computers.

Desktop computers began to appear in business in the early 1980s and originally had very little capacity, relatively speaking. Each machine could do only one task at a time. For example, one could build a financial model, or store and retrieve data, or construct a graphic presentation. But, typically one could not do all three. This turned out to be a real limitation. Simply put, one might wish to store data and later use the same data to build a financial model, and then present it graphically. But for each application the data had to be reentered. Machines simply did not have enough capacity to perform all three functions. The result was that a real business need developed to combine these functions into one application.

Within a short period, microcomputer capacity doubled and then quadrupled, then quadrupled again. Suddenly, the limitations lifted and equipment that combined financial modeling, graphics, and data manipulation into a single unit became available. Many readers are familiar with the software product Lotus 1-2-3. Not surprisingly, this product became and still is phenomenally successful. The price of the microcomputer needed to operate Lotus 1-2-3 today is the same as and sometimes less than computers with far less capability only a few short years ago.

The point is that the microcomputer migrated from being a machine that could perform a single task to one that can "integrate" several tasks. Now, if we take that development as a trend and extrapolate, a partial vision of where this particular technology is going emerges. Predictably, as users of current equipment become more proficient, they demand that more functions be integrated. These functions include word processing, data communications, and the like. Not surprisingly, Lotus 1-2-3 was followed by another software product named Symphony by the same company (Lotus Development Corporation) which integrated word processing into the basic trio: modeling, graphics, and data manipulation.

Somewhere along this migration route we will eventually encounter a desktop machine with excellent, user-friendly graphics and an electronic version of all the functions the professional or manager encounters in work: modeling, data communications, text processing, calendaring, electronic mail, graphics, and so forth. This machine may bear a striking resemblance to LISA, the Apple computer which performed poorly from a marketing standpoint. Here again, we see how technical feasibility got ahead of demand. As users become more accomplished with current technology and begin to demand more capacity, then—and only then—will the technology migrate forward.

The question is less one of *where* the technology is going as *when*. And when the migration will occur is less a technical issue than a user acceptance issue. This point is pivotal in the Integrated Technology Management Framework for it highlights user acceptance as a critical factor in managing technology. More capacity; more sophisticated applications; more demand for greater capacity. The common microcomputer is *migrating* by continuously adding other functions handled by professionals, migrating so that it becomes a sophisticated workstation, migrating so that it becomes much like Apple's LISA machine, and then migrating beyond.

But how does one predict where the migration will eventually go? The answer is found by looking at what bigger, more powerful, expensive, and sophisticated computers can do now. What can they do? They can solve extremely complex problems such as simulations and manipulate large volumes of data. They can process knowledge or symbolic information. They can produce extremely high-resolution graphics. They can recognize complex voice and image patterns. And they can do all these things extremely fast. Advanced engineering workstations, mainframes, supercomputers, parallel processors, and artificial intelligence systems do these things now. Desktop computers will do them within a decade.

Interactive Computing

The second trend that abetted distributed processing was interactive computing, which in turn was the outcome of two important developments in computing: data bases and on-line access. Both developments have their roots in the 1960s, but interactive computing came into its own during the 1970s. Interactive simply means that information can be immediately and directly retrieved. Remember that in centralized batch-mode processing, retrieving data required long turnaround times and originally involved physically taking the data to the computer to be processed.

Data bases changed all this. Data bases meant that data could be warehoused under the direct control of the computer; the computer could get to it anytime it needed to do so. And retrieval was efficient because the computer went directly to the information of interest without the complicated machinations required by earlier batch processing. There are even more fundamental implications of data base. It also permits rapid retrieval of data based on any combination of criteria of interest; the so-called what-if capability. One could, using the simplest example, ask how many customers over fifty-five years of age, earning more than $70,000 annually, and living in the Southwest bought certain products in the past six months, *assuming* that the appropriate data had been collected and stored in data base format.

Originally this issue of data collection was important. Early data base systems were extremely expensive to operate compared to batch processing. For years, debates raged in many companies (and still do) about whether to collect all data that could conceivably be used in the future or just that for which there is a present and identifiable need. The debate was often resolved by the decreasing costs of these systems. So inexpensive have these systems become that data is now stored in great quantities and on a speculative basis.

The capabilities of data bases have only been sampled. These capabilities will be vastly improved by computers designed specifically for data base processing and will again be heightened by the incorporation of advanced technologies such as artificial intelligence and parallel processing into the data storage and retrieval process. Already, two decades of data base technology have begun to revolutionize marketing. Some of the most lucrative and interesting applications of information technology today relate to how companies promote and sell their products and services. Once mundane mailing lists, for example, now have the capability to pinpoint specific customer target groups. As another example, American Airlines has found that its data base of 3.5 million frequent flyers has numerous uses in marketing, especially when interfaced with other data bases owned by hotels and rental car companies.

But what many people have missed about data base systems is that they were a *fundamental change* in information technology. The ability directly and immediately to retrieve information has created and continues to create substantial change not only in marketing but in the total change of various industry structures; in other words, in the way business is conducted. In office automation, data bases are the cornerstones of both electronic mail and electronic filing. In factory automation, information stored in data bases determines what actions robots, automated material-handling devices, and related systems will take next.

On-line access has closely accompanied the introduction of database; the two really go hand in hand. On-line access is best exemplified by the computer terminal which has become the most common device for this purpose. If data bases provided immediate and direct access to information stored in the computer, on-line capabilities provided immediate and direct access to the computer, especially by nontechnical employees. The technical computer specialist was no longer a necessary intermediary between the nontechnical user and the machine.

On-line access had profound impacts of its own. Computer use was no longer confined to the computer room—access was possible throughout the company's facilities. It was no longer tended solely by specialists but increasingly was available to numerous employees. With many people interacting with a single computer, it became a medium through which all employees could interact and communicate with each other. Increasingly, nontechnical people not only began to understand the nature of computers but began to take more and more control of the resource. Coupled with data bases, on-line access was the key development that caused computer processing to become decentralized. The ability to decentralize was the chief characteristic of *distributed* computing.

Telecommunications

The third major trend contributing to the decentralization of computer processing was the growing use of telecommunications for sending data from one location to another distant location. At the simplest level, telecommunications extended the reach of on-line access. Rather than putting a computer terminal down the hall from a computer located in New York City, the terminal could now be located in Albany, New York or Denver or Los Angeles. Not only could a terminal be connected with a computer, but that computer could be connected to other computers—large mainframes, medium-sized minicomputers, or smaller desktop computers.

The combination of these three trends—increasing power at decreasing cost, interactive computing, and telecommunications—was driven by increasing technical feasibility, business need, and economic feasibility and pushed many companies forward into the distributed processing environment. Essentially, the technology was moving closer and closer to nontechnical users within the organization, breaking down the central data processing unit's control. Distributed processing became a catch phrase for the ability to combine technological components—computers of all sizes, capacities, and costs; data bases; and telecommunications access—in endless configurations. And for managers, an understanding of this principle lies at the very heart of effective utilization of information technologies. Simply stated, the principle is this: Computer-based technologies have *migrated* from a centralized environment which limited the organization's ability to functionally fulfill its mission to a form that accommodates the natural functioning of the organization. Three important implications—concerning organizational fit, business goals, and structural change—flow from this principle.

Organizational fit

Managers are free to fit the technology to the organization rather than forcing the organization to fit the technology. Indeed research has shown that use of distributed processing is more likely with multiple, unrelated product lines and geographically dispersed, highly divisional organization structure. In other words, the more complex and diverse the organization, the more complex and diverse the technology.

As previously noted, with distributed processing many people began predicting the demise of central data processing operations. Such is not likely to be the case for most large companies that rely heavily on information technologies. Central data processing will continue to be used for processing large volumes of transactions and performing other computing tasks that require complex systems. These include maintaining corporate data bases and complex communications networks. Both tasks require significant amounts of computing power and expertise, and applications for both are growing rapidly.

BUSINESS GOALS

As information technologies migrated to the distributed mode and came increasingly under the control of nontechnical users it was inevitable that these users would bend the technology to the business interests of the firm. At Manufacturers Hanover, managers began to recognize by the mid 1980s that distributed processing had gone about as far as possible under the control of the central processing group. This provoked a shift in thinking: Technology-based information services had gone from being a tactical necessity to a strategic weapon in the institution's marketing arsenal. An official described the bank's conclusions this way:

> First, we recognized there was need for further disaggregation of operations. Not just another decentralization of facilities, but the realization that operations, as an essential resource of the banking, *needs to be under the control of the business unit* (emphasis added). Second, our vision recognized technology as a core resource of the corporation, needing to be institutionalized and fitted into the overall structure so that technological advances can be leveraged against corporate goals.

The realization also caused management at Manufacturers Hanover to restructure the business units by aligning them with their five major client categories. The five sectors, previously described, were created and, most importantly, each sector's head was given *direct* management control over technology resources. It was at this point that the central operations unit declined from 8,000 employees and a $500 million annual budget to 2,000 and $150 million respectively. Of this component about 700 were information technology professionals.

STRUCTURAL CHANGE

The third implication is that distributed processing capabilities are increasingly changing the structure of doing business. This structural change takes place at two levels: within the industry and within the company. Michael Porter has written extensively about competitive strategy and advantage, including the role of technology:

> Technological change is one of the principal drivers of competition. It plays a major role in *industry structure change* (emphasis added), as well as in creating new industries. It is also a great equalizer,

> eroding the competitive advantage of even well-entrenched firms and propelling others to the forefront. Many of today's great firms grew out of technological changes that they were able to exploit. Of all competition, technological change is among the most prominent. (1985, p. 164)

According to Porter, technology can modify industry structure in several ways:

1. Technology can change economics of scale, thus raising or lowering barriers for firms wishing to enter the industry.
2. Technology can change the bargaining relationship between the industry and its buyers or its suppliers.
3. Technology can create new products or product uses which can substitute for existing ones.
4. Technology can change the basis of rivalry, for example, pricing decisions.
5. Technology can alter industry boundaries, widening them to include firms from other industries or narrowing to focus on a particular segment.

In this book, we shall focus more on managing information technologies in *response* to competitive changes within the industry. This issue will be considered more deeply in Chapter 6.

The second level of structural change takes place within the company. Distributed processing promotes the *ability* to make changes in internal structure. Because of their almost unlimited flexibility, at least in theory, information technologies can be used to change the organization in unlimited ways. Or, to quote Porter again:

> A firm, as a collection of activities, is a collection of technologies. Technology is embodied in every value activity in a firm, and technological change can affect competition through its impact on virtually any activity. (1985, p. 166)

Ultimately, we know, as Porter tells us, that technology can be used to change industry or firm structure in numerous ways. We know also that appropriate changes can alter the economics and/or create new or different products and services. We hope that these activities will lead to a better competitive position. But in managing technology, the question eventually becomes: How can the company be changed with technology to produce the desired effect? *What can technology do to allow the company to be restructured in a more competitive way?*

Effects of Technology on Organization Structure

We shall be pursuing the answer to these questions in one way or another throughout the remainder of the book, but we begin with three primary effects technology has on organization structure:

1. *Automation* is the technique of making an apparatus, process, or system operate automatically, without human intervention.
2. *Disintermediation* is the process of removing an intermediate place, stage, or process.
3. *Integration* is the process of uniting with something else or incorporating into a larger whole. Integration occurs in many different forms.

These effects do not operate singly, but in interaction. Nor does a specific sequence exist. Integration may be more important in one period of time or stage of development, leverage in the next. Nor are these effects the exclusive domain of computer-based technologies. To some extent they have operated throughout the history of mankind. For example, primitive hatchets made of stone provided leverage that enhanced a human's effectiveness in chopping down a tree or killing an animal. Early irrigation systems and, later, mills that used the gravitational power of water contributed to both the disintermediation of human effort and automation. With the arrival of the Industrial Revolution, which created economic changes due primarily to the introduction and use of power-driven machinery, these same three generic effects were operating despite major social dislocation. Integration became a particularly important factor during this time as large-scale production replaced small-scale craft work. Currently, we tend to overlook the fact that computer-based technologies have not brought these effects as innovations but are little more than a variation on the same set of three themes. These effects have led both us and our forefathers to the same result: Technology, through its three effects, has either changed the economics of doing work or led to improved or new products and services.

Automation

Automation is fairly well understood by most readers; computer technologies are increasingly able to replace human labor in both factory and office employment. As we shall see shortly, the replacement of human labor rarely yields the most desirable benefits from technology, although during the 1960s and 1970s it did. During those two decades, as we have already seen, the bulk of transactions processing was *automated.* This formerly labor-intensive activity has now reached a mature state and fewer opportunities for direct labor-replacement gains are possible. In future office automation applications, the same is true. Although much is being said nowadays about streamlining office staff, especially middle management, this reduction can be attributed to forces other than pure automation. Even in manufacturing, automation of human labor per se is a secondary benefit, if any benefit at all. The cost of robotic equipment, including installation and maintenance, often outweighs the cost of retaining humans to do the same job. Other benefits of technology (discussed below) swing the balance.

One example is the banking industry's use of automated teller machines (ATMs). New York-based Citibank now has an electronic network of over 600 self-service terminals which handled 100 million transactions in 1986, or more than 13,000 per month for each machine. Numbers like this would seem to suggest that ATMs pay off handsomely by replacing human tellers. But limited experience to date in the industry suggests otherwise. Banks are experiencing the "33 percent wall," the point beyond which consumer acceptance seems resistant. Additional marketing dollars are then needed to promote additional ATM use which is needed to improve the economics of scale. Banks are also faced with the issue of whether to charge for the privilege of using ATMs to help offset upkeep costs.

The traditional view of technology is that it must cost justify, and from that perspective, many managers may have difficulty understanding ATM technology. But automation provides other benefits besides productivity improvement. By replacing human labor, a much greater range and consistency of performance can often be achieved. The 33 percent or so of banking consumers who accept and use ATMs may do so because teller services are offered twenty-four hours per day. But increasingly, these consumers will find that other financial services will be offered through the same machine. Perhaps the consumer will be able to invest in an IRA, buy stocks and bonds, or purchase life and auto insurance. Thus, whether the motive is improving productivity or obtaining some other benefit, *automating an activity is the first step in using technology to restructure the business.*

One point remains. Many managers have a confining view of which human activities can be automated and which cannot. For years, even now, we have clung to the idea that only routine, structured tasks can be transferred to computers. Yet information technologies have increasingly moved into higher reaches of the organization and have begun to take on tasks that were previously considered to be the domain of human experts. The question at hand for almost any activity within a company is not *whether* a computer can do it, but *when*.

Decision support systems have been the primary use of computers for advanced or sophisticated data processing applications. While there is a wide array of different types of decision support systems, one might broadly characterize these applications as the *automation* of quantitative analysis techniques. Quantitative techniques, until recently, saw little broad use in business primarily because they were not cost effective; too much human effort was required to calculate an answer when an approximation would suffice. Yet as technical and economic feasibility of distributed computing improved and business need became more sharply defined, decision support systems prospered.

A perfect example is afforded by the auditing profession. The profession espoused statistical sampling methods for decades. Yet in practice, samples of transactions were taken in a way that would have barred valid application of probability theory. True, some attempts to randomize samples took place, but truly random samples were not generally achieved. With the arrival of the microcomputer, audit support systems began to appear that could generate a truly random sample of invoice numbers, for example, based on a range of numbers and the number in the population. These numbers could then be used to draw the actual sample. Audit review procedures were performed and the resulting data entered into the system. The system would then generate the inferences needed to make judgments about whether financial records were, as CPAs say, "fairly presented." Previously, this statistics-based procedure would have required the time of an expensive professional and the added benefit of using truly random numbers would have been marginal. As a result of the trend in automating audit procedures, the demand for college courses has increased dramatically. Most universities previously only mentioned statistical sampling in their audit curriculums. Now many schools offer advanced courses. Boundless other examples of the fruitful marriage between computer technology and quantitative techniques exist.

Debates have raged for years over the question of how useful the application of quantitative techniques, automated or not, is to managerial decision making. Clearly, computer-based decision support has helped in solving complex, ill-defined, unstructured problems. But, if decision support systems have been nipping away at the domain of human knowledge-workers (i.e., managers, professionals, administrators), then a recently evolving segment of information technologies has marked this group as its primary target. That technology will do for knowledge-workers what early computer technology did for transactions processors within organizations. If the introduction of commercial computing was the first major milestone, if interactive computing which led to the distributed environment is the second, then undoubtedly the third is machine intelligence, the automation of knowledge.

Until recently, machine intelligence was almost unheard of by the business community and, in fact, only in the early 1980s did knowledge automation enter the commercial arena. But when it entered, it did so with a bang. Within a few years the number of applications has grown meteorically. Part of the allure is that machine or "artificial" intelligence (as it is often called) is not a separate, discrete element but a method or an approach that could potentially pervade all aspects of information technologies. This means that some degree of machine intelligence can be incorporated into almost any organizational process.

Artificial intelligence is actually only one of several emerging advanced technologies which include supercomputers and parallel processing. Supercomputing basically refers to the fastest computers available and, of course, the definition of "fast" changes with time. Parallel processing is more difficult to explain, but it essentially means that certain limitations in processing speed can be by-passed so that a quantum leap in computer performance results. A complex problem is broken into subunits and these smaller problems are worked on simultaneously, the results being combined at the end.

Certain aspects of these different technologies have begun to merge so that all will eventually blend into one. Taken together, these combined technologies represent, in the words of one research scientist, "a complete redesign of the computer"—the so-called fifth generation. Fifth generation systems, when they arrive, will focus on knowledge information processing as opposed to data processing.

Current artificial intelligence systems help perform a variety of "knowledge-based" tasks requiring human expertise. They configure computers, provide repair consultation for steam locomotives, interpret geologic data, control gas processing plants, manage investment portfolios, analyze corporate bad debts, consult in personal financial planning, advise on telecommunications network maintenance, and perform geographically based market survey analysis, to name a few applications. Anywhere human expertise is needed, machine intelligence will play a role.

One of the leading examples of expert systems is XCON, a system developed in a joint venture between Digital Equipment Corporation, and Carnegie-Mellon University. This system, also known as R1, designs the configuration of components in almost all computers produced by the company and performs at a level which exceeds that of human experts. Based on a customer's order, XCON develops the specific configuration of the computer system from thousands of components. A set of instructions is then generated for actual assembly of the system. Over three thousand specific rules had to be built into XCON. In this case, machine intelligence technology was used because the configuration problem was ill-defined. Despite the seemingly straightforward nature of the problem, no unique solution exists. There is no single approach for solving the total problem which, of course, becomes more complex as new components are added.

XCON provides a powerful example of the first reason why machine intelligence technology is so important. This system takes a large number of components which can be combined in a vast number of ways, and for which there is no unique solution, and coordinates the result. We find similar problems facing us in the growing number of components and the complexity of computer-integrated manufacturing, office automation, and similar systems.

These applications also have enormous potential for altering the economies of scale for manufacturing, professional, and management work, an issue considered in some depth in Chapter 5. This potential comes from the fact that, once constructed, intelligence systems can be "cloned" at low marginal cost. Thus, the economics of knowledge or "know-how" itself are altered. A second economic benefit comes from the fact that knowledge systems can "tutor" humans.

Many detractors would argue against the contention that artificial intelligence will be a major development of information technologies. But what these detractors overlook is that this technology is only in its infancy. The rather striking successes businesses have experienced over the last few years can be attributed to a few basic concepts. Yet these successes have had two effects which will allow us to harvest much greater yields from machine intelligence in the future. First, machine intelligence has helped spark research into the way humans think, which is producing insights into how to develop the technology. And these successes are now attracting private research funds from companies interested in reaping the benefits from the technology. This, of course, will speed the development process. It is also noteworthy that machine intelligence is a cornerstone of the Strategic Defense Initiative, and considerable defense-related research funds have been committed.

Technical reasons also govern the future pay-off of artificial intelligence (AI). Although exceedingly difficult to explain to someone not familiar with the technology, the issue concerns how future developments will *leverage* human knowledge versus the mere mimicry that the technology offers now. Grossly simplified, the explanation follows:

> AI is generally defined as a computer system that performs intelligent action through a process of manipulating symbols. A symbol can represent any object or relation between objects, whether the object is a molecule, a person, or a chemical processing plant. An object can be as physical as a brick or as abstract as love. Performing intelligent action means making goal-oriented decisions based on programmed knowledge. AI systems mimic human reasoning, judgement, and learning characteristics.
>
> This general definition also fits some applications of traditional computer technology. The difference is a hierarchy of five fundamental AI concepts typically not used in traditional technology. They are the following:
>
> *Symbolic representation.* Concepts, ideas, data, and information used by AI systems are represented and stored as symbols. Symbols are related in list structures that resemble decision trees. Most list structures are created by if-then rules in rule-based systems, although some take a different approach. The stored symbols and their relationships are called a domain, which is a system's body of knowledge. Currently, a domain must be narrowly defined for the system to have practical significance. Domain information is stored in a knowledge base.
>
> *Inference.* AI systems increase their knowledge through inference (e.g. Socrates is a man, men are mortal, therefore Socrates is mortal). Using logic, the system examines current information (represented symbolically) and generates new conclusions or symbol structures. Specific inference rules, collectively called the inference engine, must be developed for each AI system. Inference rules combined with the knowledge base generate the possible solutions to a problem.

> *Control of search.* An AI system confined to a tightly defined domain can easily examine all possible solutions to determine which meets a predefined goal. However, systems addressing real-world problems may produce myriad potential solutions. Therefore, the search must be limited to steer the system toward only the most promising alternatives.
>
> *Symbolic reasoning.* AI reasoning, like its human counterpart, must deal with conflicting and incomplete information. An AI system must maintain truth by resolving conflicts and should be able to explain the reasons for its conclusions.
>
> *Metaknowledge.* Metaknowledge is reasoning about reasoning. AI systems not only solve problems, but also employ metalevel reasoning to determine how it solves problems. An AI system with metaknowledge develops its own methods of reaching solutions, a process called heuristics. Most current commercial AI tools rely on only the first two concepts; symbolic representation and inference. Sophisticated control of search and symbolic reasoning are less common. Metaknowledge is rare. (Monger, 1987, pp. 5–6)

The last paragraph of this citation encapsulates the relevant point here: we have just begun to tap the potential of machine intelligence technology, and many developments, some promising extremely lucrative benefits, are yet to come. However, even in infancy, machine intelligence has already demonstrated its ample capabilities to capture (i.e., automate) human expertise in a way we never before thought possible. General Electric, for example, has developed an expert system called DELTA-1 which has encapsulated the knowledge necessary for steam locomotive repair. The experience of the best locomotive repairmen was used as the primary basis for building the system. Then, the system's ability to provide accurate and reliable advice was improved by having it interact, just as a human apprentice might, with other repairmen. The beauty of this development, of course, is that once the expense of constructing the original system is over, the knowledge can be easily duplicated.

Disintermediation

Disintermediation is a term that derives from finance and refers to the process whereby savers by-pass banks and savings and loan associations to lend their money directly to government, industry, and other borrowers. Applied to technology, it refers to the by-passing of any intermediate process—cutting out the middleman. Disintermediation is one of the most powerful effects of automation.

Consider a simple example related to airline reservations. Typically a customer contacts either the airline reservations clerk or a travel agent. In the latter case, the travel agent in turn contacts the reservations clerk. The clerk looks up alternative flights on a screen and, after back-and-forth discussions, enters the reservation. Many travel agents now have their own computer terminals which are essentially the same as those used by the airline's own reservations clerks. With a few minor adjustments, the travel agent can now directly make reservations, bill the client, and reserve seats. In other words, the need for the airline clerk's involvement has been eliminated. The by-passing of part of the reservation process constitutes disintermediation. The ability to disintermediate the reservations clerk depends on putting access to technology in the hands of the travel agent. In fact, travel agents who cannot afford airline access systems often form sharing arrangements with those who can.

Ultimately, the travel agent will be disintermediated. This could happen in a couple of different ways. First, electronic kiosks could be built and placed in busy shopping malls or office buildings. A potential traveler who approached the kiosk would answer a series of questions about budget, interests, and so forth. The kiosk may then act in the capacity of the "expert" travel agent (based on artificial intelligence technology) and advise the client about possible destinations. The kiosk could then run a video of the suggested destination much as Club Med and other resorts do at their sales offices. If interested, the traveler could make reservations, receive tickets and reservations, and charge the cost to a credit card. The electronic kiosk is really a glorified access terminal. Just as the airline reservations clerk was disintermediated by giving agents access, so can agents be disintermediated by making electronic kiosks widely available.

But, the process can be taken in step further. A second scenario is home shopping. Once consumption is, in essence, automated—that is, once access to technology is extended to each family unit through home systems—the functions described above for the electronic kiosk can be available in homes. Even the kiosk becomes superfluous.

Futuristic? Not in the least. Technically, the elements are all in place. The economics are coming into bounds and the business need is present. Some companies are already experimenting with electronic kiosks to market financial services products such as auto insurance. And ATMs—machines that most people think of as a simple replacement of the human teller—actually show the strongest promise in their ability to market and sell a broad range of financial services products: banking, investments, loans, insurance. The point here is not to paint a portrait of our future with technology but to demonstrate how the disintermediation process works.

Another example of disintermediation is middle management. The rise of mid-management ranks, especially following World War II, occurred because of senior management's voracious appetite for information owing in part to large multinational, multiproduct, multibusiness enterprises. American companies often have two to four times the number of managers as a percentage of total workforce as their leading foreign trading partners. Trimming unnecessary fat in management ranks has, of course, become a much discussed issue in recent years.

Information technology is allowing and will continue to allow senior management to maintain their appetite for information and reduce mid-management ranks responsible for data analysis. This is a disintermediation process similar to that described for airline agents; it is a simple question of giving senior managers access to the technology as travelers were in the example above. With our traveler, the technology must be made *relevant* and user-friendly by means of artificial intelligence and interactive video. The same is true for senior managers; the machine must be taught to deal with senior managers in a *relevant* and user-friendly way.

Integration

Integration, like so many other terms in computing, is used in many different contexts. But here it means little more than the combining of processes, information, or whatever. At the simplest level, the computer itself is an integrative device. Even before interactive processing during central batch-mode days, the systematization of various work processes caused them to be combined in a way that produced benefits. For example, incorporating pay rate tables into an automated payroll system allowed changes in the tax schedule to be made once, to the table, in order to revise rates for all employees. Previously, tax deductions would have to be adjusted individually on each employee's record.

Interactive processing, using data bases and telecommunications, permitted a quantum leap in integration. Before data base systems were available, computer applications were developed in relative isolation so that "isles of mechanization" were created. A perfect example is employee records. In the early days of computing, payroll and personnel systems were developed separately. This meant that the same information about a single employee had to be entered twice, once for payroll and once for personnel. This situation created obvious problems. The two sets of information often got out of sync, and when changes (e.g., address, telephone number) were needed, they had to be done twice or more. Thus, redundancy in information storage was common.

Data base systems essentially provided the capability to combine all information into a single data set. The beauty of this arrangement was that it eliminated redundancy and improved the consistency and integrity of the data in a single stroke. Now, each application—payroll, personnel—would use the same data base, taking from it only the information it needed.

The key point to be made here is that the integrative qualities of data base manifested themselves partially by *restructuring the business*. Previously, the payroll and personnel units maintained separate sets of data. A data base system could combine the two into one and became a specialized task. Indeed, a genre of technical specialists called data base managers came into being.

The ability to integrate data finds expression in an ever increasing variety of forms which are having a significantly increased impact on business practices. Merrill Lynch, the investment banking concern and member of the growing family of integrated financial service companies, created a product called Cash Management Account (CMA). CMA is largely a by-product of integration by data base. It combines previously autonomous financial services such as checking, savings, money market funds, and investment banking. Technically the product is little more than the computer's ability to cross-reference (integrate) one type of account with the others. Plus, some marketing, of course. McGraw-Hill is another company which has made extensive use of the integration concept to compete. While firms like Merrill Lynch look to computer-based integration to enhance essentially native products and services, for McGraw-Hill, use of the technology in this fashion has almost become an end unto itself. This case will be taken up in some detail in Chapter 6.

COMMUNICATIONS-BASED INTEGRATION

Integration takes on a new level of meaning with communications technology. Taken to its limits, it means that any information or process on earth can be integrated. Telecommunications has already had such a powerful impact on our society that an explanation of its influence is perhaps unnecessary. One need only think of the credit card industry's growth over the last couple of decades, for example. Telecommunications has been fundamental to the trend towards deregulation in banking. The ability of banks to combine traditional banking with other financial services depends directly on integration through telecommunications. The globalization of world financial markets is but one other variation on the same theme.

Returning to the example of ATMs and electronic kiosks, one can clearly see how integration and disintermediation are closely intertwined; they are the warp and woof of automation. Data base systems permit various services (e.g., checking, investments, insurance) to be combined; telecommunications extends (or integrates) these services throughout the country or world, giving consumers direct access to data bases, and numerous intermediaries ranging from bank teller to insurance agent are disintermediated. There is, in other words, considerable potential for restructuring the financial services industry, and that is precisely what is happening.

Progress is marked by the inevitable interplay of technical feasibility, business need, and economic feasibility. There is, however, one major stumbling block that must be dealt with: our most extensive and frequently used telecommunications network—the telephone system—is suitable for voice but not particularly well suited to computer data transmission.

The reason is largely technical. Computers are typically based on digital technology while the telephone system is largely analog. For one computer to communicate with another, digital data must be converted to a form that can be sent over an analog communications network and then once again converted to digital for the benefit of the receiving computer. If the communications network were digital, such conversions would be unnecessary. Another, more important, reason relates to the speed, accuracy, and reliability of sending information over a digital network.

There is much more to digital communications technology than simply the ability to send information more quickly, accurately, and reliably, and less expensively. Digital communications technology provides the basis to begin to combine different forms of information: data, text, sound, and image. From a computer's point of view, there is no difference between data, text, sound, and image information. Although humans regard each as belonging to a separate category, the computer ultimately converts each to a series of positive and negative electrical charges. What this means from a computing stance is that various forms of information can be stored together in a single data base. One may electronically retrieve a file on a favored artist, for example, and learn his birth and death dates (data), read about his life and accomplishments (text), hear his music (sound), and see a portrait (image) either as a still or moving picture.

We have already begun to see the beginnings of data, text, sound, and image integration. Voice synthesizers give us telephone numbers when we call the information operator. Devices exist that permit us to give voice commands to computers. For years, we have been moving closer to having a telephone that allows us to see the person to whom we are speaking. And telephones incorporating limited computer capabilities are becoming more common. The implications deriving from further integration of data, text, sound, and image are profound. This, of course, is why American telecommunications companies are pushing to install digital communications media such as fiber optic cables.

So critical is telecommunications integration to the ability to restructure business practices, and the economy, that Nippon Telegraph and Telephone Corporation (NTT) is undertaking the complete digitalization of Japan's networks over the next decade. This system, called the Information Network System (INS), will permit transmission of voice, data, and image over a single communication line, eliminating many complexities in the present communications system. Higher-speed communications will be possible at lower cost including high-speed communications links to every Japanese home.

The Stages of Integration

Currently, American progress might be called "creeping integration." Instead of a sweeping conversion to digitalization, integration is occurring in piecemeal fashion so that its final form will be realized only after slowly progressing through various stages. Six stages of integration can currently be identified:

Stage 1. Isolated Integration

Stage 2. Computer-Integrated Manufacturing

Stage 3. Office Automation

Stage 4. Company-wide (Enterprise) Integration

Stage 5. Interorganizational Integration

Stage 6. Consumer Integration

Not all stages are relevant to a given firm. Stage 2, for example, applies only to manufacturing concerns. Nor will a firm move sequentially through various stages of integration. A company may be working on Stages 3 and 5 simultaneously, while another business segment remains at Stage 1.

Stage 1. Isolated Integration

In Stage 1, integration is piecemeal and largely based on data base integration and random telecommunications links. Current technology in U.S. businesses is almost exclusively at Stage 1, because managers have tended to use technology as a response to specific problems. The weakness of this approach is that the pieces may never make sense as a whole (i.e., become integrated). Although Japanese managers also undertake piecemeal automation, their motives stand in stark contrast to that of American managers. Japanese managers begin by visualizing how the office or factory will ultimately operate using technology. They develop a blueprint of the future. Then they begin operations with largely unautomated procedures. As time passes, technology is substituted gradually but within the overall master plan. In other words, piecemeal automation is done in context.

Sanyo, for example, designed a VCR plant around this concept seven to eight years before VCR demand was strong enough to support building a totally computer-integrated plant. Production started using human labor but automation was eventually substituted. IBM and other American companies that have automated have also discovered that while automation may have to be accomplished in a number of small steps, often for financial reasons, having a firm vision of how the enterprise will appear later down the road is key.

Stage 2. Computer-Integrated Manufacturing

The second stage of integration concerns manufacturing. Because manufacturing is a complex process, it is notoriously difficult to control. Large numbers of white-collar paper pushers are typically required to keep track of materials, work-in-process, and finished goods inventories as well as deliveries from suppliers and demand for product. Quality of inputs and outputs as well as reliability of the process itself are all factors in a complex equation where numerous constraints are operable. The goal, of course, is to produce the right amount of product and do it at a steady rate. In other words, fluctuations in production and interruptions disrupt supply and cause costs to increase. Thus, coordinating all the diverse components and activities is the challenge in most manufacturing.

Historically, the predominant method of preventing work stoppage and smoothing workflow within different stages of the production process has been to buffer materials inventory. Sufficient raw materials are kept on hand to begin production and enough subassemblies are retained so that the failure of any production unit does not starve successive units of inputs needed to continue operation. Since stores of raw materials and subassemblies must be kept on hand, this idle inventory translates into an inventory holding cost which can be significant. Indeed, eliminating it is a primary objective of computer-integrated manufacturing. It is the fat in manufacturing that must be rendered from the factory. Although many people believe that reduction of labor costs is the major motive of factory automation, such is not the case. Labor is typically a small component of total manufacturing costs and direct replacement of machine for humans rarely cost justifies.

How does computer integration squeeze idle inventory out? By converting all factory equipment into computer-based technologies, it incorporates the ability to receive, store, and send information. Machines can also be programmed for various activities. Thus, lathes, drilling presses, conveyor systems become "intelligent" or programmable. They become robots and automated materials handling systems. Robots are really the desktop computers of the factory floor in that they control individual actions but also can communicate with other equipment.

This communication allows information to be passed back and forth between equipment and this information flow is controlled by a central computer (i.e., computer-*integrated* manufacturing) which coordinates production, smoothing workflow automatically. And because the computer is precise and can handle vastly larger quantities of information, it greatly reduces the inefficiencies. It is far more effective than the army of white-collar paper pushers.

But the impact of computer integration goes much deeper. The driving concept behind human-coordinated manufacturing is to commit the factory to the production of a single product or set of products for as long as possible. Information technologies permit a different approach. Because robots and other equipment can be reprogrammed quickly and control is automatic, "retooling" can occur within a very short period and is very economical. A single plant can be programmed to handle a whole range of products within certain design specifications simply by entering instructions in the central integrating computer. This computer then gives appropriate retooling and reconfiguration commands to all robots, materials handling systems, and other equipment.

> Manufacturing used to be based on the assumption that the most important operating dynamic was economics of scale—i.e., large numbers of identical components could be most economically manufactured on automated equipment. Long production runs and rigid product (and process) specifications were essential to low cost production, market share, and profits. Change was avoided because it was expensive. This argument makes sense when downtime for tooling changes stretches into weeks or when equipment must be replaced to permit product change. (Jelinek and Goldhar, 1984, p. 32)

Thus certain aspects of new computer-based technologies are based on the "economies of scope" rather than economies of scale. It may be as cheap or cheaper to produce a variety of products or different numbers of products on the same machine. Not only can products be tailor-made to closely match the demands of a particular market niche, but small order lots will be economical as well.

> "One off" production is no longer fetched: the economic order quantity approaches one with electronic controls, increasing responsiveness to both market change and customer requests. Inventory holding becomes a needless expense if replacement parts (even for older designs) can be easily scheduled without disrupting operations. (Jelinek and Goldhar, 1984, p. 33)

Stage 3. Office Automation

Office automation has for most managers been the greatest point of personal contact with information technologies. Although central data processing has certainly been a presence for many years, it has been a distant presence at best. Day-to-day interactions have been largely confined to receiving computer-generated data or in fewer cases use of the computer terminal. The most tangible evidence of office automation, as distinct from data processing, has been the localized use of word processing and personal computers, although many companies have gone much farther. Voice messaging is becoming more commonplace. As local area networks grow so does use of electronic mail and calendaring. Teleconferencing is gaining support. Though use of office automation is growing rapidly, applications are piecemeal. As we have already seen, even major companies lag behind in attempting to apply its full capabilities, and when managers have tried, they have often been disappointed at the level of performance attained.

Thinking on office automation has followed very closely the pattern created by central data processing: a problem or opportunity is defined, and the technology used to provide a solution. But we are finding that this piecemeal approach does not work effectively in office automation. Nowhere is this more evident than when managers evaluate the benefits of giving a professional employee a desktop computer. Ostensibly, the goal is to "improve productivity." But managers find they are hard-pressed to "cost justify"—cost justification (i.e., discounted cash savings, payback, etc.) has been a standard tool for decision making for past computing applications. As legions of managers will attest, measuring the benefits when technology is applied to office workers is not easy. Measurable benefits are often paltry. Some managers proceed "on faith" that the investments will be worthwhile in the end. Some rationalize. Some hesitate.

Confusion exists because we have used the wrong paradigm for thinking about office automation. But if a carryover in thinking from central data processing is not appropriate, where are we to obtain a useful paradigm? The answer, though it may seem obtuse at first, is from computer-integrated manufacturing systems (CIMS) discussed in Stage 2.

Conceptually, no difference exists between computer-integrated manufacturing and office automation. In fact, we could refer to automated administrative processes as computer-integrated office systems (CIOS). But the manufacturing analogy is easier for people to grasp because it produces tangible results and the process steps are more routine and structured. Offices involve intangible processes—communication, ideas, concepts—and outputs are sometimes difficult to identify. Indeed, the failure to articulate the end products of office work may be responsible for more inefficiency and slack in corporations than many managers commonly recognize, a problem that goes far beyond the issue of technology but one that is being brought to a head by technology.

Adapting CIMS thinking to the office provides some startling conclusions and gives solid direction with respect to how office automation should be managed. First, managers must completely purge the office of the applications-oriented approach to automation. Instead, the objective should be to design and build an administrative "factory" using computer-integrated manufacturing as the role model. Just as an automated manufacturing system can produce a range of products that fall within certain constraints (IBM's Lexington plant can manufacture almost any product in a 26″ × 22″ × 18″ work envelope), so must the office design be flexible enough to handle a range of administrative activities. The object is to design an administrative factory that can be reprogrammed to produce any number of end products that share some common constraints.

Suppose, for example, that rather than adding office automation equipment to an existing environment, the slate were wiped clean and a completely new automated office could be designed and built from scratch. To some degree, Toshiba Corporation was able to do just that when it built its new office building in Shibaura, Tokyo. The focal point of the building construction was "optimum effectiveness in OA (office automation) applications." In other words, Toshiba built a flexible office factory based on extensive research, with one thousand units of office automation equipment, a fiber optic local area network, and more.

For most managers, windows of opportunity to build an automated office factory from the ground up rarely present themselves. And, one might argue, when this does happen, these opportunities are rarely recognized. But remember that even though Sanyo's VCR factory was designed and built for total automation, early stages of utilization depended on manual labor. By the same token, office managers must design the totally automated office and then put pieces of the puzzle into place over time.

Why must the office be redesigned for total automation? What benefits can be expected from tackling the problem from this perspective rather than the current approach of automating selected applications? Again, computer-integrated manufacturing provides the example. As we have already seen, most manufacturing costs are related to materials, not labor. Thus, the greatest economic benefit to be derived is eliminating excess inventory from the system (disintermediation) through vastly improved (computer) coordination.

This reduces the cycle time (or responsiveness) between the customer's order and delivery. The thrust comes not from replacing labor per se but in coordinating the elements of production, squeezing the inventory fat out of the system.

In the office environment, the pattern of resource expenditures is not unlike production, in that much is wasted on coordination. One representative study (Poppel, 1982), showed that knowledge-workers spent only 29 percent of their time on intellectual work such as reading, analyzing, and document creation. Almost half the time was spent in meetings, varying from 40 percent for nonmanagerial professionals to more than 60 percent for senior managers. Another 25 percent was spent on less productive activities such as travel to meetings, waiting for a machine to become available, seeking information, expediting assigned work, and performing clerical tasks. Taken together, meetings and less productive activities accounted for 71 percent of a knowledge-worker's time. This is where the lucrative benefits of office automation lie; not in the intellectual activities, although automation of these tasks is both possible and desirable.

To say that communications and information retrieval should be automated is nothing more than saying that the primary technologies (communications, data base, small-scale accessibility) should be applied. But it must be painfully obvious that in order to achieve their benefits of interest—the disintermediation of the communications and information retrieval processes—the entire office must be integrated. With integration, the time devoted to contacting and communicating with other people, the time wasted waiting for a machine to become available, the time spent searching for information, and the time needed to supervise other layers of employees are all drastically reduced. And this attacks 71 percent of the total time expenditure by knowledge-workers, by far the most lucrative portion.

I am suggesting that time is to administrative processes what inventory is to production. Inventory buffers the inefficiencies in the unautomated factory. If the fat in manufacturing is inventory, then the fat in offices is time. Time buffers the inefficiencies of individual components and prevents the systems from coming to a standstill. Telephone tag is one of the most celebrated forms of these inefficiencies.

In IBM's Lexington plant a manager commented that one could judge how successful a computer-integrated plant was by watching for materials "on the floor." What he meant was that any production materials that were on the floor of the plant were not incorporated into the system. A complete system, one that is totally integrated, would not create excess material to sit idle on the floor even for a short time. This exact thinking must be carried into our mentality about office automation.

This can be accomplished by recognizing that the key to achieving office automation benefits is to keep all inventory off the floor and in the system. In the office this simply means that someone who is outside the system is "on the floor." If the similarities between factory and office automation are valid, then manufacturing may provide some valuable clues about the effects of integration on productivity. Recall that manufacturing automation does not hold much promise of reducing labor costs directly by automating jobs piecemeal. Labor savings are expected to occur only with complete integration. Transformation involves changing the fundamental nature of manufacturing even to the point of redesigning the product. Real cost reductions are predicted to come from the reduction of inventory.

We have already argued that significant productivity gains in the office will not come from automating labor directly. First, integration must occur, and the key to integration is communications. Reaching this critical level of integration will permit the transformation of office work. At that point the nature of office work will change even to the point of redesigning the work product.

Stage 4. Company-wide (Enterprise) Integration

As the automated factory and office take shape, the next logical level of integration will be company-wide: between factory and office. Many companies have already developed some electronic ties between the two. What areas can benefit from company-wide integration? The first is marketing and sales. By being able to directly interrogate the computers that control production, salespeople can determine delivery schedules based on current production demand. When an order is taken it is entered directly into the computer and automatically scheduled for production, with a scheduled delivery date. Before production begins, the computer orders the necessary parts and materials from vendors, advising them of required delivery times. Thus, those who are responsible for purchasing are also tied into the factory computers. And, because all inventory is in the physical manufacturing system and the computers automatically track its progress, information for accounting purposes is already in the machine.

Far too many managers regard such scenarios as futuristic, interesting but not practical. Yet, while visiting one of the most advanced computer-integrated manufacturing facilities in the United States, I asked plant managers what comments they heard most frequently from senior managers who came to observe. Their response was that the senior executives were most surprised by the extent of integration in the plant. In fact, the plant's owner had made the plant operational with standard commonly available hardware (i.e., robots, automated materials handling systems, etc.), although some software development was necessary.

In order to realize the potential of the computer-integrated factory, this firm had to push somewhat beyond what was currently being done in manufacturing. But now that the trail has been blazed and others will follow, the natural question becomes "What is the next frontier?" The answer is company-wide integration. Thus, managers who view manufacturing automation and office automation as separate concerns are creating modern-day "isles of automation" that will hinder company-wide efforts to achieve integration.

Stage 5. Interorganizational Integration

Technology does not make a distinction, legal or otherwise, where companies are concerned. Thus if employees within the company can access data remotely through the telecommunications network, then so can people outside the company if they are provided with the technical capability. In other words, there is no technical reason why someone outside the company should not be able to get into the company's system; by the same token, someone within the company could get into another organization's computer system. This means that the integration and disintermediation process does not stop with a single organization. By extending the computer system to parties that share some interest with the organization, additional benefits can be obtained. As we shall also see in Chapter 5, interorganizational integration may greatly enhance the company's ability to compete using technology.

Computer systems that cross organizational boundaries are not new. A primary use of such systems in the past has been the formation of linkages with customers. One of the most common examples is airlines. Airlines have for years provided various travel agents with computer terminals that allow the agents direct access to their data bases. By doing so the airlines have not only set up an interorganization network but they have also disintermediated the airline employees that would have otherwise been required to handle seat reservations.

Insurance companies have also long been users of interorganizational systems. By providing both company-owned and independent agents with computer terminals, direct access to rating rules and other pertinent information has been possible. Again, one can quickly see the disintermediation of the elimination of insurance company employees who were formerly required to provide the information personally that is now provided by the computer directly. Another fairly well-known example is American Hospital Supply. This Connecticut-based firm provided hospitals with computer terminals to directly access their order system data bases. Instead of having a sales person call on each hospital to take orders for medical supplies, hospital staff could now enter orders directly.

Another direction for growth in interorganizational systems is suppliers. If one thinks back to the computer-integrated manufacturing system which controls all aspects of the manufacturing process, one can see computer-based communications with suppliers as a logical extension. In other words, there is no need for human intervention in the process of translating production requirements into orders from vendors and the placing of those orders. Eventual integration of the factory with suppliers is important. To function smoothly the automated plant must be continuously in motion, and a disruption in the flow of supplies would be disruptive to the entire process.

Interorganizational systems ae expanding in a number of other areas as well. For example, links to corporate data bases are increasingly common for auditors, financial institutions, and others. These links pose a number of problems. For example, issues such as which organization will be responsible for entering data, maintaining telecommunications networks, and developing software must be resolved. Legal and quality control issues are also at stake.

Closely related cousins of interorganizational systems are public access data bases. Generally speaking, these data bases provide information. Common examples include Dow Jones New Retrieval, Dun & Bradstreet, and McGraw-Hill. As the variety of these services grows, corporations are likely to find their use more cost effective. As a result, there will be an integration of these outside electronic services with the internal operations of the company resulting in further disintermediation.

Stage 6. Consumer Integration

Just as electronic communication does not stop at the company's borders, neither does it stop at simply including other organizations. The communications-driven integration phenomenon extends all the way to the end of the line, so to speak, to the individual consumer. Computers have been an important factor in boosting the current direct marketing movement. Direct marketing simply means any activity in which individual prospects are identified by name, as opposed to other forms of marketing in which consumers may be known as a group but not as individuals. A special version of direct marketing is telemarketing. Telemarketing has become big business but does not always depend on integration using communications technology. In many cases, telemarketing simply involves using a computerized mailing list, credit cards, a toll-free telephone number, and access to credit card approval data bases. In telemarketing, disintermediation occurs because of the elimination of the retail outlet. In other words, a direct link exists between the wholesale company and the consumer. Consumer-oriented technology has begun to appear in many different forms and this trend is likely to continue.

We have long held a preconception that the human being is an important component in many transactions with customers. We are now finding to an increasing extent that this is not the case. Examples previously cited—electronic kiosks and ATMs—demonstrate to some degree that much of the consumer's dealings with companies can be automated. Some complicated transactions may in fact require human advice in the process of execution. But with machine intelligence developing so rapidly, even the bastion of human involvement is rapidly eroding. In other words, computers are increasingly likely to provide expert advice along with executing the transaction.

The electronic kiosks are of course small islets of electronic technology that as of this writing have not been connected via communications to the central corporate information systems. But such connections will inevitably be made for kiosks located in high-traffic, high-density shopping districts. And they will be made in another way as well: by electronically linking the consumer in his or her home to a network that provides two-way access between the company and the consumer. This phenomenon most readers will readily know as "home shopping." What this means is that even the electronic kiosks and similar devices will one day themselves be disintermediated. Why should a consumer interested in purchasing an insurance policy, investing in a particular company's stock, researching and purchasing an airline ticket, bother with going to a remotely located electronic facility when one is available in the home?

Extension of technology to the consumer is simply another variation on the themes played over and over in this chapter: automation, integration, and disintermediation. As a result, it will undoubtedly alter the relationship between the company, the customer, and the intermediaries (i.e., agents). This in turn will change the structure of the industry and the fundamental basis of competition.

Variations on Themes

Assessment requires that managers develop a sense of the short- and long-term possibilities of information technologies before moving on to specific questions about impact on industry and strategic uses. I would forcefully argue that any senior manager today who does not understand these possibilities is woefully unprepared for his or her responsibilities. This is not a question of technical detail but of understanding the major forces that drive information technologies and the nature of technology development. This chapter has been an attempt to identify these forces.

Managers should take courage from the knowledge that computer technology has evolved so that it can be fitted to the organization's mission. The days of central processing that often subverted organizational processes are pretty much at an end. Although the technology has become more complex and manifests itself in seemingly endless ways, that diversity gives managers the freedom to mold the technology to the organization. Indeed, a theme that is repeated throughout the book is that organization follows mission, technology follows organization.

Managers should also see forward movement with technology as an interaction between technical and economic feasibility, and need. Technical capability does not in itself create momentum. As specific technologies take root they begin to affect individual organizations, industries, and even societies by automating, disintermediating, and integrating. Each process feeds on the other. Looking for the potential for these forces to affect the business is the starting place for thinking about using information technologies strategically. But before this thinking can take place, managers must systematize their understanding of what the effects will be. This is the process of assessing technology and the subject matter of the next chapter.

4

Assessing Strategic Opportunities

Seizing Initiative

Far too often, managers in companies which could benefit competitively by being more aggressive with technology wait until the software or hardware vendor shows up on the doorstep before considering a new technology. Yet by the time this technology is so broadly available, it is highly unlikely that any company could use it to create a significant, sustained competitive advantage. That action simply comes too far downstream in the course of the technology's development to be of much value.

Part of American management's weakness in assessing new technologies can be attributed to the piecemeal approach to applications taken by most managers. This posture has been largely reactive: computer technology is called into play largely in response to a problem or business opportunity. Computers have also been used to enhance productivity by streamlining transactions handling. More recently, managers have begun to realize that technology is not simply a tool, available for the wanting, but a powerful force that is reshaping the competitive forces within their industry. Suddenly, thinking has shifted. Managers are now beginning to see that control over technology often leverages the company's competitive position, while failure to control technology means losing it. Now that the stakes have increased, pressure has mounted to be much more involved in reconnaissance of new and emerging technologies. And where technology plays a key role in business strategy, the company should be correspondingly aggressive. Aggressive means thorough and active assessment of new and emerging technologies. Aggressive means working with vendors—IBM, Digital Equipment Corporation, and others—where appropriate to influence the course of technological development. Aggressive means putting research and development dollars behind technologies that may be key to the company's particular competitive strategy. Aggressive means having in-house capabilities to monitor, analyze, and evaluate new developments.

Companies that effectively manage information technologies display that aggression. The Travelers, for example, maintains a relatively large staff of computer scientists whose function is to *know* new technologies, to evaluate them, to influence the course of their development when feasible and, when not, to redesign the technology to suit the company's needs. As a result, it is not unusual to find software created by Travelers specialists which foreshadows the general market by two years or so. Travelers has also been extensively involved in joint ventures which are developing and applying expert systems technology to the financial services business.

A second reason why managers have been slow in developing technology assessment expertise relates to the lack of a coherent U.S. industrial policy, an issue considered more fully in Chapter 9. In Japan, for example, active efforts to coordinate economic development insure that the ramifications of new and emerging technologies are carefully considered. This fact is reflected in commitments such as Nippon Telegraph and Telephone's digitalization of the nation's telecommunications infrastructure by the mid 1990s. Another example from Japan is the so-called fifth generation project which seeks to develop machine intelligence. Coordinated policy is also the driving force to convert the telephone to an information retrieval device in all French homes. In the United States, the only comparable effort is the process of developing technology for defense purposes. Currently, the primary example is the Strategic Defense Initiative (SDI), or "Star Wars" program. While there are undisputably commercial benefits that derive from this type of research spending, experts agree that the commercial potential derived from each R&D dollar spent is small when compared with efforts, such as those in Japan, that have direct commercialization as the goal.

Current Thinking on Technology Assessment

Some companies, such as General Electric, Exxon, and GTE, have technology assessment units or individuals with assessment responsibility. According to the Fordham survey, about 70 percent are senior-level positions. But only about 20 percent of the companies responding reported having a senior-level position with technology assessment as the primary responsibility. Most of the people in these positions perform other duties as well and are therefore not totally focused on technology assessment. Over half of the technology assessment positions were vice-presidents, senior vice-presidents, or executive vice-presidents, although in isolated cases the president and even chairman were identified. But, most of these titles indicated that the position was responsible for information services or resources. Others included "new developments," R&D, and corporate planning. Twenty percent said a manager or director with a technical title (e.g., director of advanced technology, director of engineering) was the chief technology assessor. Planners, internal consultants, and evaluators were also mentioned but much less frequently.

This data implies that companies do, generally speaking, attempt some form of technology assessment. But, assessment is typically one of several responsibilities and is more often than not aligned with the information services or another technical unit. Individuals who are at a high level and come from a business planning perspective are still relatively scarce in practice. Recognition of the importance of technology assessment as an independent function with a business planning (i.e., nontechnical) perspective is on the rise, however.

Many individuals currently engaged in technology assessment have a technical background from science and engineering. But business experience is also considered important recognizing the need to bridge the gap between the technical and nontechnical worlds. And generally, the person filling the assessment role should have years of experience as opposed to being newly hired. In 1986, the *New York Times* reported on these qualifications in an article entitled "Corporate Technical Assessment." The article stated that breadth of knowledge was more important then indepth technical expertise noting that:

> Such a director probably will not read scientific journals so much as *Scientific American* and other popular technical magazines. The director of technology assessment also attends meetings to keep up with trends and is aware of new patents, and the new director uses "outside resources" such as contacts with universities known for expertise in fields the company is interested in. (Fowler, 1986, p.21)

Three points from this excerpt are worth noting. First, it is exceedingly difficult to imagine that a major corporation, especially one in which technology is key to competition, would rely upon *Scientific American* and other popular technical magazines as a substantive source of information about new and emerging technologies. Second, keeping in touch with universities who are key players sounds much more reasonable but hardly sufficient when one considers the wellsprings of technological innovation in the country. Here one suspects that the desirable background in science or engineering may be desirable because of familiarity with the research and development establishment: the whole panoply of military and civilian research agencies, universities, incubated high-tech start-ups, large research-oriented corporations. Third, technology management as it is viewed from a scientific/engineering perspective may have important methodologies (for assessing new and emerging technologies, for example) that should be explicitly incorporated into general management thought now that technology has become a general management concern.

As it turns out, a well-developed technology assessment methodology does exist. And while this methodology focuses on social and political impacts of the introduction of new technologies, when recast for commercial assessment it provides an *excellent* starting place for thinking about the potential impacts of a particular technology on an industry's structure. This in turn provides the foundations for informed strategic thinking at the corporate level.

Technology assessment is not a well-understood term, nor for that matter, a well-understood process. Historically, technology assessment studies were typically undertaken to examine the social, economic, and political implications of a new or emerging technology:

> . . . a class of policy studies which systematically examine the effects on society that may occur when a technology is introduced, extended or modified. It emphasizes those consequences that are unintended, indirect or delayed. (Porter et al., 1982, p. 3)

Although technology assessment seen from this perspective is only rarely of direct concern to businesses, methodologies used for these purposes offer insight into how a technology will develop and what its impacts will be. This process can be adapted to a commercial perspective because it offers insight into where new business opportunities are likely to be created and where existing business activities are likely to become less profitable and perhaps obsolete. Thus, technology assessment is a tool for understanding the overall business environment. Just as market research is conducted to understand consumer preferences, for example, technology assessment should be conducted to understand technology.

How can a company establish a means of importing reliable and appropriate information about outside technological innovations? How are technologists within the company reoriented from a system of largely internal, inbred communication to one that increases reconnaissance with outside technological developments? How does a company transform itself from passive recipient—almost victim—of technological change to active participant—to making change happen in a way that contributes to the companys competitive strategy?

Management should establish the organizational mechanisms necessary for performing the appropriate level of technology assessment within the company, keeping six objectives in mind:

1. *Survey new and developing technologies*. The first order of business for the assessment unit is to establish and maintain an ongoing understanding of all information technologies. The required knowledge does not have to be deep for every aspect of technology but it should be a basic working knowledge.

2. *Assess the commercial potential of the technology*. As it becomes obvious that a technology can make a significant contribution to the company's competitive position, the company should naturally deepen and strengthen its assessment efforts to determine the technology's commercial potential. Could the company become a vendor and/or a user of the technology?

3. *Assess impact on the industry*. General knowledge of technology or knowledge of specific developments must then be applied to the question of what the impact on industry structure is likely to be. Will technology make existing products or services obsolete, or create new ones? Will the basis for rivalry change? Here we are dealing with the classic "chicken-and-the-egg" question. Does technology drive business goals or do business goals drive the technology? No

simple answer exists, of course. We can emphatically say, however, that the issue should be tackled from both sides of the fence. As new generations of managers who can more effectively bridge the gap between the technology and business perspective come along, things will improve.

4. *Educate management*. Once those responsible for assessment conclude that a particular technology will have an impact on the industry, senior management must be brought on board and brought up to speed. The age when management can avoid taking a "hands-on" approach to technology is over. And management "involvement" or support is not enough. Especially in situations where the impact on the industry is significant, managers must provide the leadership with respect to the employment of technology. It is unreasonable, of course, to expect senior managers on their own initiative routinely and thoroughly to brief themselves on new technology developments. But, some regular procedure is needed to update senior managers about relevant technologies. This is where technology assessment comes in.

5. *Assess impact on the firm*. With an understanding of the technology itself, a sense of its commercial potential, analysis of its impact on the industry, and an informed management, those responsible for technology assessment are free to turn their attention to what it means for the company. The objective at this stage is to identify alternative postures the company could take in light of changes in the technology *and* the business environment. A subtle but salient point resides here. As seasoned business planners will know, strategy is rarely laid in response to a single variable in the environment. In all industries, other forces are at work. In financial services, for example, the impact of technology must be considered against deregulation and globalization of world financial markets.

6. *Decide on involvement with new and emerging technologies.* At some point, management must take a position with respect to a technology that is having an impact on the industry. The position may derive from an active decision to invest in a new technology or not. A de facto position may be the result if management takes no action. Whatever the method, a position is taken and it sets the course for competition in the new business environment. The issue here is that those responsible for technology assessment have a role to play in position-taking. This role will assume greater dimensions in cases where technology is important.

Assessment Levels

Management will undoubtedly wish to examine in some depth a technology that is likely to create substantial changes in its industry. By the same token, a technology with only incidental significance will receive much less scrutiny. Resources committed to assessment should, of course, correspond to the task.

Experts have outlined five levels of technology assessment; macroassessment, miniassessment, microassessment, monitoring, and evaluation. Each level represents a different magnitude of commitment.

1. *Macroassessment*. A macro-level assessment involves a full-scale study of the technology. This means a thorough investigation of every aspect of the technology and how it might have a bearing on the business. Quantifying the commitment of resources needed for this level of assessment is, as with all levels, not possible before the objectives of the assessment are specified. The resources needed vary over time and funding of the technology assessment function is in itself a key decision. A macro-level assessment requires major funding; it would be typical to expend five to ten person-years, for example, to accomplish the effort. Certainly when one considers assessment of the role of computer-integrated manufacturing technologies in the auto industry, involvement is total. The same is true for telecommunications technology as it relates to the airline or financial services industries.

Macroassessment by its very scope is broad, and depth is therefore not the objective. Macroassessment sets the stage for major involvement with a particular technology for which a decision has already been made in most cases. Thus, macroassessment is likely to be followed by a smaller, more narrowly focused study at some point.

2. *Miniassessment.* As its name implies, miniassessment is smaller in scale than macroassessment and fulfills one of two major purposes. First, like macroassessment, it can be broad but shallow. Examining the potential for expert systems technology in the insurance business might be one example. Or the assessment could be narrow and in-depth. For example, a company might exhaustively examine expert systems capabilities to provide personal financial counseling to customers as a lead-in to selling insurance, investments, and other financial services products.

3. *Microassessment*. A micro-level assessment is again smaller in magnitude than a miniassessment. A rule of thumb might be one or

two person-months of effort. Little in the way of concrete results or recommendations can be expected. Microassessment is a quick look into a new technology to determine the scope of the subject and the key issues. Those responsible for technology assessment may conduct a microassessment by attending seminars by leading consultants, making fact-finding trips, talking with vendors and customers. Microassessment is distinguished from monitoring (see below) only in the sense that microassessment is a discrete one-time investigation.

4. *Monitoring*. Monitoring is an ongoing assessment of a technology. Often a prior assessment activity has identified a need to systematically follow a technology's development. Monitoring is appropriate when one or more of the three factors creating forward movement, discussed in Chapter 3, is absent. A technology might be technically feasible with a business need, but not economically ripe for exploitation. For example, for years Exxon routinely monitored point-of-sale technology. It wanted to install an electronic appoval system to eliminate the need for employees at its retail outlets to telephone the company for approval of large credit card purchases. The problem was that the cost of telecommunications, a key component required to make the system work, remained too high to make the technology feasible for a period of time. But by monitoring these costs over time, Exxon finally discovered that telecommunications costs had declined to the point where the system was feasible and shortly thereafter it was installed.

5. *Evaluation*. The purpose of evaluation is to provide ongoing feedback on previous technology assessments. If earlier assessments prove to be incorrect because of changes in the course of the technology's development or because of inaccuracy in the prediction process, a reevaluation of plans is obviously needed. While the course of technology is much more predictable than most American managers commonly believe, technology forecasting is still an imperfect process. Thus, careful ongoing assessment of progress is useful in effectively managing technology.

The difference between the five assessment levels is not simply one of magnitude. Each serves a specific purpose: solving a specific problem, ongoing surveillance, or gaining a general understanding. Macro, mini, and microassessments are generally undertaken as discrete projects, whereas monitoring and evaluation tend to be ongoing activities. The assessment function's structure should reflect the difference in these various levels.

The Assessor's Role

The person responsible for technology assessment has three distinct, albeit overlapping, roles: intelligence gatherer, "gatekeeper," and planner. The assessor therefore is a bridge between technical people, business planners, and operational managers within the company. The assessor is also an all-important link with sources outside the company. How these various roles are played out will depend on the role technology plays in the company's competitive position. If technology is relatively unimportant, a single assessor with other assigned responsibilities may be adequate. Where technology is key to competition, the rank of the assessment function should correspond to the complexity and magnitude of the task.

As intelligence gatherer, the assessor is a major conduit for information about new and emerging technologies. Generally speaking, the more important technology becomes to the company the farther upstream the company's assessor should go to examine the development process. At a minimun, a company must reconnoiter what the industry and competitors are doing with any particular technology. Attending conferences and seminars, reading *Scientific American* and other journals, and reviewing current technology products available from vendors may be a start. But these activities do not constitute a program for technology assessment. In other words, a company is highly unlikely to create, much less sustain, competitive advantage with a technology that is common knowledge. If information is commonly available, competitors could employ it as well.

Moving farther upstream—closer to the wellspring of innovation for a particular technology—can take several different forms; including developing closer relationships with vendors, undertaking sponsored research or joint ventures, incubating upstream companies, and internal innovation.

Closer Relationships with Vendors

Many major technology vendors such as IBM and Digital Equipment Corporation (DEC) maintain close ties with companies that are already or have the potential to become significant customers, or that are doing leading-edge development work that might spur commercial use. These vendors typically consult this favored group of customers on product design and other key issues. It is therefore possible to have some degree of influence in their product development, though in the final analysis this tends to be minimal.

Some companies can serve as beta sites or early delivery locations for new equipment. A company that serves as a beta site is essentially helping perform a full-scale, live production test of the new equipment, the advantage being the opportunity to observe the equipment well in advance of the mass market. An early delivery arrangement allows a company to be one of the first to receive the equipment once shipment of a new product begins. Often, price reductions and delays in payment are given to a customer willing to be the first to use the equipment in production. In some cases, a vendor may enter into a joint venture to develop commercial applications of its new equipment.

While each of these programs moves the company further upstream, the specific characteristics and timing of the technology remain firmly under the vendor's control and rarely is the arrangement exclusive. The greatest value comes from receiving information in advance of other competitors, a benefit that may or may not have significant value. Again, the eve of a new equipment line's introduction may be too late to be of competitive value.

Sponsored Research

Companies may also move upstream in the technology development process by becoming active financial partners in university research. Digital Equipment Corporation, for example, sponsored research on artificial intelligence at Carnegie-Mellon University, a leader in this area, and ultimately came away with expert systems software that was applied internally at DEC and later marketed to customers. Various universities are noted for strengths in one area or another: artificial intelligence, telecommunications, supercomputers, robotics. These universities have often developed their strength through federal funding from the Department of Defense, the National Science Foundation, and other government sources, or through corporate support.

The paradigm of how well this arrangement can work is the Media Lab at the Massachusetts Institute of Technology (MIT) located in Cambridge. The Media Lab was originally formed as an association of academics from different backgrounds—art, psychology, education—who shared a common interest in computer technology. Populated by leading scholars and staffed by bright, creative reseachers, the Lab soon developed a reputation for innovative applications of technology. Recent work includes three-dimensional holography targeted for use in automobile design, for example. Another experiment involves artificial intelligence-driven corporate newsletters which incorporate newspaper items in a format structured to individual preferences. The system might learn, for example, that the first item of interest for a particular user, in order of priority, is yesterday's sales figures and the second is the Mets baseball scores. Still another example includes "movies of the future," stored on CD-ROM disks rather than videotape.

Many of these developments may at first seem obtuse and unrelated to the practical concerns of business. But two important facts make the Media Lab relevant to how companies compete with technology. First, the Media Lab is an influential institution when it comes to shaping the technology of tomorrow. In other words, they are *further upstream*. Second, the technology they are shaping in media is largely at the behest of major corporations. Indeed, millions in funding come from many of the nation's top corporations, at a time when the technology is still in its formative stages, still able to be shaped for specific commercial purposes.

University research has tremendous potential, and the fact that the system is underutilized greatly undermines efficient technology development in the United States, a subject addressed in Chapter 9. For the company that supports research in this environment, the advantages can include:

Access to Expertise. Few companies can afford to develop the expertise that can be concentrated in university research mills. Even if a company can afford such facilities, certain economies realized by using outside personnel who have already developed their expertise are lost. The university can also provide a neutral meeting ground for other experts and thus facilitate a constructive exchange of information, an issue we shall return to shortly.

Creative Climate. Large companies focused on the daily task of earning a profit are usually not as successful in creating an innovative and creative climate necessary to generate forward thinking in areas such as technology. Sometimes an academic environment can act as a magnet for innovators and provide them with a supportive culture. Sometimes, not always. Most managers are aware of the difficulty of sustaining an entrepreneurial spirit within the company. The university research lab offers a safe haven from corporate cultures which may work against the project's goals.

"Off-Shore" Risky Investments. Sponsored research at universities offers a unique benefit over internal corporate research. Should a project not achieve its objectives, the work generated will still be valuable as a mechanism for educating students and advancing academic knowledge. In other words, the project also has an educational benefit.

Building Strong University R&D. A final advantage is that financing upstream technological developments through universities builds national economic strength. In the words of the 1985 report of the President's Commission on Industrial Competitiveness:

> University revenues do not cover the rising cost of research, and engineering faculty salaries do not compete with those of private industry. As a result, fully one-tenth of the nation's engineering faculty positions are currently vacant. In critical fields like electrical engineering and computer science, some universities report half of their positions as unfilled. As the American Electronics Association stated so well, we are "eating our own seed corn" by failing to produce the faculty required to teach future scientists and engineers. Our scanty harvest shows the result; Japan provides more engineers than we do. (p. 20)

Joint Ventures or Ownership

Much upstream technology is developed in private companies, of course. One option is to embark upon the business of creating technologies where synergy between that activity and the company's business exists. This option can be developed through joint ventures or ownership in leading technology concerns.

Apex, for example, is a joint venture between The Travelers Companies and Integrated Resources, Inc. Its purpose is to develop expert systems technology for use in Travelers' financial services business. One such product, Proplan, provides automated personal financial consultation.

Ownership provides distinctive advantages of its own. General Motors has used this approach extensively in attempting to upgrade its ability to apply technology to manufacturing. It acquired stakes in Electronic Data Systems which specialized in applications development and Tecknowledge acquisition, in particular gave it access to upstream developments in an advance technology.

Incubation

In the absence of acceptable relationships or existing commercial ventures, another option to move upstream is to incubate companies that might eventually be of service in developing technologies that would contribute to the company's objectives.

> To incubate fledgling companies implied an ability or desire to maintain some kinds of prescribed and controlled conditions favorable to the development of new firms. The incubator seeks to give form and substance—that is, structure and credibility—to start-up emerging ventures. Consequently, a new business incubator is a facility in the maintenance of controlled conditions to assist in the cultivation of new companies. (Smilor and Gill, 1986, p. 1)

For a variety of reasons, employees in large companies who want to pursue technology projects with commercial potential cannot be accommodated within the corporate structure or culture. Companies like Exxon, for example, long ago found that helping these people start their own companies was not only a profitable investment in many cases but helped the original company accomplish its broader purposes with technology. To some extent, the incubator model combines the best elements of university and internal research. It permits the creative entrepreneurial spirit to thrive while providing the venture with adequate resources and managerial expertise.

Internal Innovation

A final option is to develop technology within the confines of the company. Moving the company upstream in this case means increasing its involvement with innovation. In recent years, interest in the subject of corporate innovation, especially technological innovation, has been growing. But recent research suggests that American managers see less need for innovation within their companies than do Japanese managers, for example. Fewer also believe that innovation can make a specific contribution to corporate earnings within the foreseeable future. One of the major barriers to effective innovation is the preoccupation with short-term performance.

But many managers believe that innovation is a manageable process internally and offers significant opportunities to make the company more competitive in the technology arena. Successfully managing large-scale innovation within major companies involves at least nine "patterns" of behavior, according to Quinn (1979). These include:

1. A strong incentive for successful development
2. A clearly defined need specified in economic/technical terms
3. Multiple competing approaches encouraged at both the basic research and development levels
4. User guidance and participation to insure that specifications remain current and that the system will be used
5. High expertise and research discipline maintained by assembling first-rate people and providing them with knowledgeable leadership
6. Time horizons that are typically larger for successful innovation
7. Committed champions which carry major developments forward
8. Top-level risk-taking support
9. Clear objectives generally accepted as worthwhile

The Insularity Problem

One of the factors that weighs heavily on a firm's decision to invest in technological innovation is the problem of insularity. Disturbing evidence exists that American companies tend to be far too inbred with respect to the importation of new knowledge about technological developments and innovation in general. The Fordham survey, for example, suggests that technology assessors do not cast their net very far when gathering intelligence. Respondents identified internal company research and industry group meetings as the most useful sources of information. Professional society and conference meetings and publications ranked as being less so. Consulting firms and universities and colleges trailed the list and were rated as being only marginally important. Other research reveals that this is in marked contrast to Japanese methods of introducing innovative information. Internal company and industry sources play relatively less role for the Japanese than they do for Americans, whereas the academic community and other outside sources play a much greater role in introducing innovation, invention, and technology assessment.

Research in recent years has shed some light on why this information is so inbred within U.S. companies. The reason relates to basic differences between science and technology which develop relatively independently from one another and in very different ways.

> Despite the long-held belief in a continuous progression from basic research through applied research to development, empirical investigation had found little support for such a situation. It is becoming generally accepted that technology builds upon itself and advances quite independently of any link with the scientific frontier, and often without any necessity for an understanding of the basic science which underlies it. (Allen et al., 1978, p. 48)

Thus, contrary to popular misconception, science does not directly spawn most technological development. Technology is a separate process in which most development feeds from previous technological developments. The process of developing a technology may hit a stumbling block or problem which requires scientific research to find a solution. But, generally speaking, scientific R&D procedures do not lead to good technology applications.

The importance of this distinction is that information about scientific and technological developments is disseminated differently. Scientists typically gather and organize information and distribute the same in journals and at academic conferences. Sharing research results is an important tradition in scientific thought, which is not to say that some scientists do not keep information secret when doing so is to their advantage.

But technologists employed by commercial enterprises are typically discouraged from sharing information. First, information on technological developments has competitive value and companies obviously wish to protect their investment. Secondly, technologists by definition are applications oriented. Their mission, as employees of commercial enterprises, is not to advance the frontiers of knowledge but to accomplish specific business objectives. There is, therefore, no reason to share.

> Information is transferred in technology primarily through personal contact. Even in this, however, the technologist differs from the scientist. Scientists working at the frontier of a particular specialty know each other and associate together in what Derek Price had called "invisible colleges." They keep track of one another's work through visits, seminars, and small invitational conferences, supplemented by an informal exchange of material (before publica-

> tion). Technologists, on the other hand, keep abreast of their field by close association with co-workers in their own organizations. They are limited in forming invisible colleges by the imposition of organizational barriers. (Allen et al., 1978, p. 40)

Companies wishing to protect technology-related trade secrets are faced with a dilemma. If people are the best carriers of technological information, then the flow of that information is increased, as the movement of people between organizations is increased or as turnover increases. Most firms discourage turnover, of course. The effect of this approach may be counterproductive. Most managers believe that

> . . . a low level of turnover could be seriously damaging to the interests of the organization. Actually, however, quite the opposite is true. A certain amount of turnover may not be only desirable but absolutely essential to the survival of a technical organization. . . . (Allen et al., 1978, p. 43)

This position, taken by researchers who examined the issue of how technological information is exchanged, still leaves companies as exporters of valuable innovation as well as importers. But, more importantly, the whole process of exchanging technological ideas through employee turnover is costly and inefficient at best. At worst, it can be disastrous when the price is the loss of key competitive information.

The Japanese have handled the problem of disseminating information using a completely different approach, well known to many American managers by now. They have established a series of quasi-government agencies which coordinate specific projects related to technological developments. These agencies gather researchers from industry or subcontract directly to industry. Although the Japanese government typically retains the patent rights, firms are aware of relevant technological information and have the option to license the technology for commercial development. Researchers then return to the company's laboratories to make of it what they will. The whole system strikes most Americans as being too regulated.

Since the human carrier is the best vehicle for technology transfer, turnover contributes to the dissemination of technological knowledge between firms. The high-tech firms located in Palo Alto and along Route 128 in Boston have notoriously high rates of turnover, for example.

From a management perspective, the Japanese have long recognized this problem and have developed a system of research and information dissemination that has several advantages. The resources of numerous companies are pooled as a means of leveraging R&D dollars rather than replicating efforts from company to company, and all organizations have equal or at least equitable opportunity to use the results to create firm-level competitive advantage. The result is highly synergistic, of course. The whole industry, and even the economy, is raised up. Essentially, each firm achieves a higher plateau of technological development and knowledge. This accounts for much of Japan's success in the international marketplace.

U.S. businesses go in the opposite direction. Within the framework of free enterprise, efforts to develop technology are at the firm level and therefore the synergy—the combined efforts—produced by the Japanese system is lacking. The Achilles' Heel of the American system is individualism or, more specifically, the dislike of government involvement.

In summary, the technology assessor's role cannot simply be one of surveying the landscape and gathering intelligence for internal consumption. Instead, he or she must be actively engaged in positioning the company correctly in terms of how active it will be in innovating with technology.

Technology Transfer

No matter how sophisticated a company is technologically, the problem of transferring specific expert knowledge to business units where it can be profitably employed is a daunting task. Even companies like IBM occasionally suffer from the-cobbler's-children-have-no-shoes syndrome: their own employees are not using technology produced by the company. Transferring this knowledge from intelligence sources to in-house company experts to business unit managers is the gatekeeper role.

Traditionally, the gatekeeper function has been performed by a technical person who in many cases is in first-line supervision, not management. Research shows that gatekeepers are easily identified by others in the organization. But increasingly the importance of the gatekeeper role in management ranks is growing. The role of gatekeeper in management is different—not technical—because it focuses not only on the introduction of innovative technological concepts to the company but also on the policies and systems by which innovation will take root and prosper. The success of innovation within a company does not rest with the gatekeeper, of course, but with the way the broader scope of management activities is related to innovation and change.

Because each company's situation may differ somewhat, recommending a single method of transferring technology-related knowledge is impossible. However, Figure 4–1 depicts one company's approach. This company has extensive experience managing sophisticated defense and energy related technologies and its management understands the frustrations of keeping current with new technological developments and transferring that knowledge to people in the organization who are in a position to put the technology to work.

The company solved the issue of keeping current by organizing "technology centers" under its technology assessment function. Each center is the focal point of a specific technology, application of technology, or area of support. In Figure 4–1 examples of specific technologies include local area networks, telecommunications,

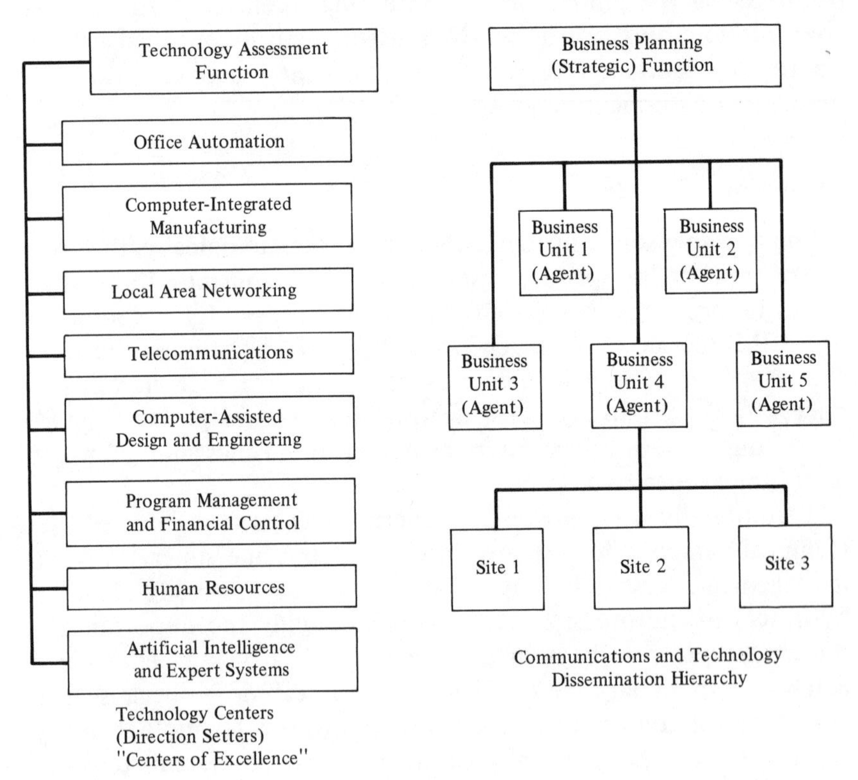

Figure 4–1. Assessment and Transfer Structure.

computer-assisted design (CAD) and engineering, and artificial intelligence and expert systems. Applications include office automation and computer-integrated manufacturing systems (CIMS). Support includes program management, financial control, and human resources. The rapid introduction of new technologies may, for example, create a special need for help in the area of managing program implementation and evaluating the financial desirability of implementing the new technology. Or the need to reevaluate hiring practices and training programs for employees in the face of a new technology may create a need for special expertise in human resources management. The technology centers are mostly direction setters. They develop superior expertise in their area of focus and are therefore able to provide quality advice throughout the company. One company refers to them as "centers of excellence."

As the commitment of resources to technology assessment is periodically evaluated, some technology centers may be disbanded if they no longer serve a useful purpose. Therefore, the number of technology centers will change over time, as will their purposes. In a real sense, technology centers mirror changes in technology itself. As technology evolves, so do the centers. The existence of technology centers does not mean that others in the company find uses for them or, for that matter, even understand the issues. There is a separate need to communicate and disseminate information developed by the technology centers to other functions within the company. The mechanism used to do this is shown on the right-hand side of Figure 4–1.

The company represented in Figure 4–1 has formed a team of technology transfer agents assigned to particular business units. The figure depicts five of these units. A specific unit may or may not have a transfer agent. If it is important, transfer agents may be assigned to a geographical site, a division, or whatever. Taken together, all technology transfer agents represent a "shadow" hierarchy that may be a rough approximation of the functional and operational hierarchy. The agents may even be organized formally into a technology planning council or similar body. The council would provide a mechanism for transfer agents to communicate their experiences across traditional organizational boundaries.

The key to the structure is that transfer agents work with functional and operational business units to understand their business and where it is going. In turn, they work with the technology centers to understand new and emerging technologies, consulting each as it seems appropriate. The assessment and transfer structure represented in Figure 4–1 is an adaption of the matrix form of organization since there are lateral as well as function divisions of responsibility. But this and similar structures can be useful to bridge the technologists' perspective and that of the business planners.

A Technology Assessment Methodology

Just as the structure for technology assessment and transfer will vary between companies, so will the procedure. However, Figure 4–2 presents a general technology assessment methodology; a framework that individual companies can adapt to their specific needs.

The methodology is based on fourteen specific components:

1. Defining the business issues
2. Describing the business context
3. Describing the technology
4. Forecasting the technology
5. Forecasting the impact on industry

Discovering Potential

Business Problem-Oriented Perspective | *Technology-Oriented Perspective*

Defining the Business Issues | Describing the Business Context | Describing the Technology | Forecasting the Technology

Predicting Industry Impact

Forecasting the Impact on Industry | Identifying Changes in Industry Structure | Analyzing Changes in Industry Structure | Evaluating the Impact on the Industry

Exploring Potential Responses

Identifying Responses Available to the Company | Analyzing Alternative Responses | Evaluating Alternative Responses

Employing the Results

Communicating the Results | Transferring the Knowledge | Participating in Position-Taking and Policy Decisions

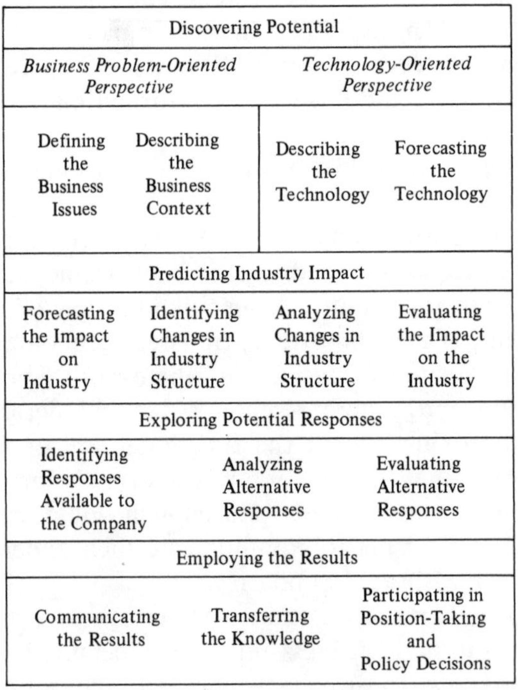

Figure 4–2. A Technology Assessment Methodology.

6. Identifying responses available to the company
7. Analyzing changes in industry structure
8. Evaluating the impact on the industry
9. Identifying responses available to the company
10. Analyzing alternative responses
11. Evaluating alternative responses
12. Communicating the results
13. Transferring the knowledge
14. Participating in position-taking and policy formulation

As Figure 4–2 shows, technology assessment can be driven from either a business problem-oriented or technology-oriented perspective. Either perspective inevitably involves all fourteen components.

The problem-oriented perspective begins with concern about a specific aspect of the business. This problem may then lead to a broader consideration of the business context and, in turn, to technology. From a technology perspective, the focus may first be on describing and forecasting the technology. But inevitably specific business issues and the larger business context must be drawn into the thinking.

The interplay between the business and technology is the fertile ground on which technology-based competition is born. The interplay may produce nothing. But it may produce an understanding that existing products or services and current methods of doing business are becoming obsolete. It may help the company to develop insight into potential new products and services, different ways of doing business, and new economic forces. Certainly a visionary management team will be the catalyst by which the insights occur. Some companies discover that their insight leads them to conclude that they should leave one business and get into another. Companies such as Singer and Westinghouse, for example, have made a transition from consumer products to advanced (e.g., defense) technologies.

As potential is uncovered by examining the relevancy of the business and technology, the methodology begins homing in on issues of more concrete concern to the company. The focus shifts from discovering potential to predicting industry impact: How will the basis for competition change within the industry? Chapter 6 presents a full-scale exploration of how technology may be used to change competition. As we shall see, several critical issues emerge at this phase of the analysis. For example, is technology (or a particular technology) at all important to the industry's future? If a technology is important, what is the timing of its probable impact? How will the industry change? Which players will stay and which will go?

As alternative scenarios of the industry's evolution emerge, the focus again shifts, this time to the task of positioning the company. Should the company take an offensive posture with technology or not? If it adopts a defensive posture, what should be done as preparation? The emphasis at this stage and throughput the assessment methodology is not to develop strategy but to explore alternatives. As a complete picture of the alternatives emerges—in terms of discovering the technology's potential impact on the business, predicting the impact on the industry, and exploring potential steps the company could take to be competitive in the future—the stage is set for employing the results of the assessment in the business.

Following is a more complete presentation of the fourteen components listed above.

Defining the Business Issues

To be realistic, any assessment study, regardless of level, must be focused on an objective. The process of defining the business issues involves making decisions about their breadth and depth, and that in turn is likely to be based on the availability of resourses. Thus, defining business issues determines the nature and scope of the study. There are, of course, numerous ways to contain a study and focus it. The study might look at only one segment of the business or business unit. It may deal with a specific time horizon—only five years out, for example—or it may be focused on business in one geographic location—securities trading in the U.S. as opposed to global markets.

The study may be bounded in other ways as well. An assessment conducted for the purpose of information for the future is different from one conducted for the purpose of near-term decision making. Unfortunately, too many companies wait until they are confronted with serious problems before responding. Immediate problems create pressure for quick (and dirty) solutions, and assessment of any variety is viewed as an obstacle.

It is incorrect to say that defining business issues as a component of technology assessment only works with long-range problems, those identified before they occur. But without a doubt lead time is important. It is important first because to the extent that the technology assessment function is mired in short-term "fire fighting," it cannot fulfill its primary mission which is to get out in front of current developments. Technology assessment is the scout of the enterprise, always traveling well in advance of the main group. A second reason, and a theme we shall return to repeatedly throughout this book, is that considerable lead time is almost always necessary to derive the most lucrative benefits of technology. It takes time to put together technology systems before pay-offs begin to occur, an issue explored more fully in Chapter 5.

Describing the Business Context

Any business problem must be evaluated in its larger context. To understand problems faced by a particular airline, one must first consider the deregulation of the industry. The same is true of financial services. Other changes might include the nature of the competition—for example, the extent to which foreign firms have become a factor. Changing consumer preferences, the economic climate—all exigencies the company must deal with in the larger scheme must be brought to light when the likelihood exists that they will be factors in technology-related decisions.

Technology assessment, if wisely done, will consider social and political factors where relevant. One large company headquartered in a medium-sized city where any layoffs inevitably generate considerable negative publicity is careful to consider how implementation of new technology systems will effect worker attitudes and unemployment. These factors ultimately affect certain decisions concerning the automation of its workforce. It does not help that the company is consumer-oriented and is therefore sensitive to public criticism.

Describing the Technology

The study must identify the boundaries of the technology to be assessed. Office automation, for example, is such a broad issue that nothing short of a macroassessment would make any real sense. The study can be focused on specific aspects of technology in a number of ways. One of the key aspects of describing a new system is the route by which the technology is developed and eventually commercialized. Differences in delivery systems will affect the degree of control a company has over the technology.

Forecasting the Technology

The purpose of forecasting, to use the words of more than one executive, is to "chart the future" or "create a vision" of what technology will be like and more importantly what it can accomplish. In the Fordham survey, managers who responded frequently mentioned that senior management often fails to see the possibilities of technology. In the words of one respondent, the greatest frustration faced in managing the technology was "helping senior management visualize the potential changes that could or would impact our environment and industry over the next five years." Likewise the manager of one of the nation's most advanced computer-integrated manufacturing plants commented that senior executives who visit often express surprise at the technological sophistication of the facility. Most believed that things "had not progressed so far."

But technology forecasting concerns not only creating a vision of the future but providing realistic information about the timing of predicted developments. What major uncertainties are there in the technology's development? What breakthroughs are needed? What potential substitutions are likely to follow? And how rapidly will the technology be diffused?

Forecasting the Impact on Industry

At this point, the assessment study moves away from the general business environment and technology to consider the potential effects on the company's industry. A forecast of industry impact addresses the question of whether the technology is important to the industry: will it make a difference?

Some technologies will have an impact on a particular industry; others obviously will not. A financial services company is not likely to commit significant resource to robotics, for example. Yet a technology such as telecommunications cuts across all boundaries; few firms will remain unaffected by future developments in this area. Thus, the objective of this component of the methodology is to determine whether the technology will have an impact.

Identifying Changes in Industry Structure

If the industry impact investigation forecasts no impact, then the assessment need go no further. But if an impact is forecast, then the relevant question becomes identifying the changes in the industry structure. Essentially, we are asking if technology will change the basis for competition and if so, how it might change. Careful identification of changes is important because they may not always be positive. A competitor may wish to be careful about setting technology-based changes in motion if they will not enhance the firm's competitive position nor sustain it.

The procedure for identifying changes in industry structure is to logically work through the consequences of introducing a specific technology. The changes may occur in a number of ways, and for certain technologies identifying these changes can be a considerable task. Multiple scenarios may exist. Thus, the emphasis is on generating all feasible scenarios. Evaluation of the likelihood of each comes later.

As an example, a financial services company develops an electronic kiosk and places it in selected supermarkets or shopping centers. The company offers integrated financial service products: insurance, investment banking, and so forth. The most visible feature of this kiosk is a monitor or screen. A potential customer can select one of several options offered on the screen. For example. the customer might ask about IRA investment options or insurance rates on an automobile. The kiosk responds with the appropriate information, even providing a printed copy for the customer. The customer might come away with a rate quote for insurance on the family station wagon.

The kiosk could go further in its service. In the course of its interaction with the potential customer, it could request name and address. This information could then be made available to a local agent who would make a follow-up sales call. Or perhaps the kiosk could make the sale by having the customer enter a credit card number. The information would be relayed to the financial services company which would generate the automobile policy and send it to the customer.

The technology behind this scenario is deceptively simple and inexpensive. Basically, it combines interactive video disk technology with a microcomputer. The consequences for competition might look something like the following:

Electronic Kiosks—Direct Marketing Scenario

1st Order. Kiosk provides identifiable marketing presence for financial services company in retail locations.

2nd Order. Potential customers can readily obtain information about company's products and services without direct involvement by company's employees.

3rd Order. Information is gathered from potential customers. Company's sales agents can focus on most likely prospects, increasing effectiveness.

4th Order. Customer has reinforcing experience with company by conveniently obtaining product or services. Loyalty to company for all financial services is cultivated.

5th Order. Sales volume increases. Costs of product and service delivery are reduced.

Later, in Chapter 6, we will look at the various ways in which companies can use technology to compete. We will further examine the electronic kiosk example from that perspective. The process of identifying changes in industry structure is essentially one of identifying alternative scenarios. Some of these scenarios will lead to a dead-end, competitively speaking. But some will offer the company the beginning of a plan to use technology to compete.

Analyzing Changes in Industry Structure

After a list of alternate scenarios for the industry has been generated, assessors must focus on the question "So what?" Or, to put it another way, this component is concerned with synthesizing a most likely scenario from the list of scenarios previously identified. The quality of the analysis will, of course, depend on a number of factors such as the thoroughness and accuracy of previous steps in the assessment process and the insightfulness of the people conducting the analysis.

At this point, managers must begin considering another critical issue which will ultimately affect the position taken with respect to the technology: To what extent will the company have to be an actor in making the scenario happen as opposed to being a passive recipient. In other words, if technology-driven changes in industry structure are likely, who will take the lead in ushering them in? The issue of using technology on the offensive versus remaining in a defensive posture will be taken up again in Chapter 6. But clearly by this point in the assessment, information needed to decide this issue should be actively developed.

Evaluating the Impact on the Industry

This is the fourth and last component related to predicting industry impact on the technology assessment methodology. By this point, assessors should have some idea about whether an impact will occur, the various ways in which the industry changes might evolve, and the likelihood of specific alternatives.

The process of evaluating the impact of a technology is one of asking what it really means for the industry: Is the predicted impact realistic? For example, with the growing use of videotex in the future, will home shopping replace the physical retail outlet? Or will it simple supplement these? Overriding this issue is whether or not the predicted change is desirable. What will be gained *by the industry* as a whole if firms begin investing in a particular technology?

The evaluation also includes looking at the players and attempting to determine who is likely to take what actions. In New York City, for example, commercial banking has traditionally been dominated by several large, powerful banks. When automated tellers were first introduced, predictably it was the large banks that could afford to finance the development costs. At first these banks assumed that smaller competitors would have difficulty following because of high start-up costs. As it turned out, smaller banks, sensing that automated tellers were indeed important, formed automated teller networks such as NYCE and Cirrus which served several banks. A customer of any bank on the network could use any of its teller machines. Thus, the smaller banks succeeded in breaking or at least neutralizing the potential stranglehold of larger banks. Whether or not the banks foresaw this particular sequence of events as the industry moved into the technology, the fact remains that they should have.

The evaluation component completes the industry-level assessment. At this point those responsible for the assessment and those responsible for strategic planning should have keenly developed ideas about different options open to the *industry*. The next major task is to explore potential responses by the company. This task has three components: identifying responses available to the company, analyzing alternative responses, and evaluating alternative responses.

Identifying Responses Available to the Company

At this point, we have not yet taken up the issue of decision making; we are still concerned with technology assessment. Identifying available responses is therefore not a question of strategic choice, since that comes later in the position-taking phase. At this juncture, those responsible for technology assessment are reacting to the industry-level portion of the assessment by asking what options are open to the firm.

The key question, of course, is whether the company will undertake to be the technology leader in the field. That decision, according to competitive strategist Michael Porter, depends on whether the costs of being the first to use a technology permit the company to create a competitive advantage and sustain it.

> The notion of technological leadership is relatively clear—a firm seeks to be the first to introduce technological changes that support its generic strategy. Sometimes all firms that are not leaders are viewed as technological followers, including firms that disregard technological change altogether. Technological followership should be a conscious and active strategy in which a firm explicitly chooses not to be the first on innovations. (Porter, 1985, p. 181)

The advantage and disadvantages of technological leadership are discussed in more depth in Chapter 6. At this point, the assessment should focus on examining leadership or followership as possible responses to industry movement.

Analyzing Alternative Responses

Once those responsible for technology assessment have identified potential responses to industry change, the question becomes what would be required to implement each response. In other words, where is the company now? Where would it have to end up? What is required (e.g., capital, time) to get there? Usually a first step is to establish a "baseline," an understanding of the company's current level of technology and related capabilities and how they were achieved.

Companies that are effectively managing technology tend to establish a clear understanding of their current computer infrastructure before formulating strategy. Often the process of establishing the baseline produces discoveries that strongly influence its technology-related decisions. One financial services company, preparing to develop a new and more responsive information system, altered course when it realized it already had $200 million invested in computer software.

Analyzing alternative responses also means raising the question of what would be needed to implement a particular alternative. Here, those responsible for assessment cannot solely concern themselves with technology per se, but must look farther to the organizational aspects of the company. Those include issues of culture, structure, management style, employee retraining, and education. Essentially, they are not only asking how the technology will change but how the organization (i.e., the company) must change. We will explore this issue more fully throughout the remainder of the book.

Evaluating Alternative Responses

The company still has not actually taken a position at this point, but now those responsible for assessment must narrow the range of options available to the company to those that are realistic. This third and final step in exploring potential responses provides the information needed for management to enter the position-taking phase.

Communicating the Results

The technology assessment methodology now shifts to the last set of components which focus on how those responsible for technology assessment employ the results of their work to date. The first component is communicating the results.

There is an important distinction here. One might say that the task is to report the results, but that would be in error. Someone can stand up in front of an audience and describe a new technology and its potential. He can do that twice, thrice, perhaps a number of times and still not be communicating. Communicating demands that the chief technology assessor work with senior management slowly, if needed, bit by bit bringing them on board with the new technology and perhaps also with new ways of doing business. Communicating involves maintaining an open conduit through which information about the changing technological environment flows in a meaningful way.

Communication about technological developments should not consist of the "one-time" approach of briefing management. When assessment is done properly, management becomes accustomed to hearing about developments continuously. Information is presented in easily assimilated form rather than in large indigestible chunks. There is recognition that time is needed to absorb new concepts. Management is immersed—kept involved.

Transferring Knowledge

Perhaps a fine line exists between communication of assessment results and transfer of knowledge. Communication is a question of keeping managers up to date. Transferring knowledge focuses on providing managers with the conceptual tools and skills they need to work with a new technology. Thus, knowledge transfer becomes more an issue as the company embraces a technology. Those responsible for assessment essentially have the task of ensuring that the management team thoroughly understands the technology for the simple reason that the better the understanding, the greater the probability that appropriate decisions about strategy and implementation will be made.

Participating in Position-Taking and Policy Decisions

Finally, those who are responsible for technology assessment must have an ongoing involvement with all phases of technology management. As we shall see in position-taking, the information technology strategy chosen may require a keen awareness of future trends, or it may not. The knowledge gained through assessment is also important in policy formulation, for it helps determine what organizational changes are needed. This latter point is extremely important since understanding the timing of technological changes can be synchronized at least to some extent with needed organizational changes.

Conclusion

Technology assessment is a two-way street. The bulk of our discussion has concerned researching new and emerging technologies, determining their impact on industry and the firm, and importing that information. The size and complexity of the unit responsible will reflect the importance of technology in general to the company and, within narrower time periods, its importance at any one time. Where technology plays a pivotal role in the company's competitive health, assessment should also flow the other way: from company to technology. By this, I simply mean that management may want to consider reaching outside the company to actively control the course of technological development.

A degree of control over a technology may have significant advantages. Control may give a company a lead in technology-based competition or may allow the company to influence changes in industry structure. This stance is clearly visible at the Japanese communications firm, Nippon Telephone & Telegraph, which has undertaken to install a digital communications network throughout Japan by 1994. It is illustrated by joint ventures, such as DEC's project with Carnegie-Mellon University researchers to develop artificial intelligence tools, and it is evident in companies such as The Travelers which have worked with major computer vendors to influence their product design. When these firms and others like them see the need for technological development in areas that will be of use to their business strategy, they reach out and do something about it.

PHASE II

Position-Taking

5

The New Economics

Three Stories

In the early 1970s, I was just beginning my career working as a computer systems professional in Exxon's retail credit card operations. My job largely consisted of evaluating applications for possible automation and during this period I performed numerous cost-benefit justifications. Projects lived and died on the basis of my net present-value calculations. It seemed possible to quantify any technology decision with little effort.

At the time, Exxon employees received a discount on purchases of gasoline and other products. This discount was deducted from the amount owing on the credit card account and a special system was required to administer these benefits. I was asked to evaluate the benefits system to see in what ways it could be improved. When I examined the large discounts being given to employees I could find no *quantifiable* benefit to offset the cost. Based on a strict application of cost-benefit analysis, I therefore recommended to management that these benefits be eliminated. My recommendation caused much mirth among my immediate superiors and was the occasion of much ribbing thereafter.

In my naivete, I had carried the quantification process to the extreme, allowing it to replace common sense. Yet years later I often reflected that I was not the only person who failed to apply good judgment. As we saw in Chapter 3, the emphasis on capturing measurable productivity gains in transactions processing during the early days of central computing caused many managers to lose a degree of control over their business priorities. Quantification of benefits was the law.

Near the end of the decade, when I was living and teaching in Boston, I undertook a consulting assignment for a mid-sized insurance company. Their Methods Department wanted to standardize discounted cash flow techniques for evaluating computer projects. Previously, they had been using payback and other similar methods. Privately, I was amused that this otherwise successful company was so unsophisticated in techniques that had been so routine at Exxon.

Those feelings were reinforced at one meeting in particular which included the manager of the Methods Department. As an aside, he asked me to provide some gratuitous advice. The company was thinking of setting up a large computer as an "information resource." Anyone could come use it for any purpose. Desktop computers had not yet made an impact on business practice and "end-user computing" as we know it today did not exist. So, the idea was still somewhat novel. "Would it be reasonable," he asked, "to allocate the cost at the end of each month based on the quantity utilized by each department?" I replied no, that in the early phases of coaxing people to experiment with the resource, volume of usage would be low and allocations of total cost would be very high. This would discourage use. The company would either have to provide the computer as a free resource for a while or charge a nominal usage fee. These recommendations of course meant that the cost of the computer would remain largely unallocated.

Consternation showed in the manager's face. He then rephrased his question and I rephrased my answer. We repeated this process but I clearly was not giving him what he wanted to hear. On the way out of the meeting, I asked another of the company's managers what was behind the question. He laughed and said that the company's president had recently bought a new model mainframe computer over a golf game with the computer's marketing representative and this machine was sitting downstairs on the loading dock. "He's got to justify its existence!" The incident further fueled my thoughts about how unsophisticated the company was in using quantitative evaluation techniques. Yet when I reflected later on the experience I realized that the president had made a sage decision in acquiring the machine. Larger issues were at stake than the ability to quantify the benefits.

A final story: More recently, I was having lunch with the president of a large American aerospace and defense company. The president was no stranger to sophisticated information technologies but during our discussion it became obvious that he was greatly concerned with economic justification issues. Five years before, he pointed out, computer-related technology totaled only 5 to 10 percent of the company's capital expenditures. Now, that amount was over half. The company's management had begun to rethink how to better manage this resource and measure its effects. But, the president said, the issue might not be *how* to measure, but *whether* to measure benefits. He then proceeded to relate the following story.

On a recent business junket to Japan, he had occasion to share a traditional tea ceremony with a highly placed Japanese executive. The American recalled thinking that this Japanese executive was probably "beating his pants off" in competition, but he decided to ask his question anyway: "How do you justify expenditures on office automation for professionals when benefits are difficult to quantify?" The Japanese executive laughed and said, "Don't you know when you have a good idea? Don't you trust your judgment that giving a desktop computer to a professional employee will improve the business?"

The moral of these three stories is this: Our current understanding of technology's *economic* consequences is wholly inadequate. *Economic* is the operative word here. It includes financial issues managers have become accustomed to associating with computer technology: cost-benefit analysis, payback period, and the like. But it goes further by encompassing the whole range of activities associated with the production, distribution, and consumption of goods and services. Economics includes economies of scale, marginal utility, marginal cost, and similar concepts. Time horizons become a critical factor in economic thinking. How long will it take to reach a certain economy of scale, for example, and once it is reached how will it affect the marginal benefit derived from the technology?

Information technologies have now become complex enough so that a sophisticated understanding of their *economic* behavior and impacts is needed. For reasons that I will argue in this chapter, technology has outgrown the simple financial performance measures now in common use. Our continuing attempts to evaluate costs and benefits using these techniques is distorting our decision making; they should be only one factor in mesuring current technology. Yet if we are to design new methods of performance measurement and management control, we must first understand their economic basis. In other words, they must be based on economic realities, rather than what we assume to be the economic performance of technology. Indeed, we must challenge the notion, as the Japanese businessman in the third cautionary tale did, that measurement is needed at all.

Surprisingly, no well-developed and articulated body of management knowledge exists with respect to the economics of information technologies. Research agendas, management curriculums, and practitioner journals are all curiously silent on the issue; contributions are meager. There is a great irony here, for it was the explosive costs and disappointing performance of computing facilities that goaded many managers into becoming involved with the technology in the first place. Otherwise, these managers would have been largely content to leave the whole thing to the technical people. But during the 1960s and especially in the 1970s, data processing budgets soared, sometimes experiencing 30 percent annual growth rates so that annual increments were often equal to the entire budget only a few years before. In the mid 1970s, one author attempted to calm management's fears with a model of data processing growth. Among other things the model suggested that there were times when increases in computing costs would exceed the rate of increase in sales and other times when they would be less. The model implied that some equilibrium would be reached in the final "maturity" phase. Everything would right itself in the end.

In fact, many companies entered something of a mature phase in their central data processing operations and costs did behave in this fashion. But a new element had been introduced: office automation. Desktop computers had appeared on the scene and their use was escalating during the early 1980s. Suddenly, the experts were saying that all knowledge-workers (i.e., managers, administrators, professionals) needed to be equipped with a personal system. Increasingly, emphasis was being placed on automating a segment of employees who made up more than half the total workforce; the white-collar workers. While the three to four thousand dollar price tag for each desktop computer seemed digestible, in quantity the cost of this equipment was staggering. Large companies found that they were spending tens and hundreds of millions of dollars on these machines with no end in sight. The prospect of replacing a handful of large mainframe computers periodically now seemed child's play compared to the thought of having to replace thousands of desktop machines. Thus, the pressure on managers to make decisions about expenditures on technology stayed alive and well.

Many frustrations still remain with central data processing equipment: costs are high, difficult to predict, and benefits fail to meet expectations. With the advent of desktop computers and office automation, problems in accurately forecasting costs were as notoriously difficult as ever but now a new factor was added. It was *virtually impossible* to measure many of the benefits. When managers or professionals were given their own machines, no measurable reduction in the size of the workforce took place and the other quantifiable advantages were scarce. In the Fordham survey, the concern about whether technology decisions resulted in realizable benefits was a major frustration identified by respondents:

> Identifying, quantifying, and tracking productivity benefits realizable through the application of new technology in management support systems is difficult to justify given the substantial cost.

Managers often proceed with white-collar automation "on faith," assuming that benefits will eventually emerge. But this faith has been difficult to sustain as the tens and hundreds of millions in expenditures mount up and investors are demanding acceptable returns. It is difficult to maintain in the face of one particular bone-jarring fact: since the late 1960s the American economy has not seen one significant rise in aggregate white-collar productivity despite billions spent on office automation. Or, to put it another way, the rise in productivity has been by far disproportionately small compared to expenditures. Managers who invest heavily in this technology are likely to experience the same plight as the farmer who drove to market without realizing that his watermelons had fallen off the truck along the road. He had nothing to show for his efforts once he reached town.

What is *wrong* with our current means of financially evaluating information technologies? Must we proceed on faith or is the economic behavior of technology predictable and therefore measurable? The remainder of this chapter focuses on two major themes in response to those questions: the recent shift in economic behavior and the resulting conflict between economic behavior and current measurement standards.

SHIFT IN ECONOMIC BEHAVIOR

The fundamental economic behavior of information technologies has changed. During earlier days, computers were introduced on a piecemeal basis, primarily to automate lower-level clerical activities such as transactions processing. This processing was itself piecemeal. The systems analysis and design phase focused on breaking down larger processes into smaller processes, individual components, or steps to determine which could be automated and which must continue to be performed by human beings. The emphasis was on applying the automation capability of the *computer*, thought of pretty much as a stand-alone machine with limited influence. Because applications tended to be discrete sets of activities, evaluation methods such as cost-benefit analysis were probably appropriate for the early years of information technologies.

But, an underlying shift began to occur and we did not fully comprehend its implications. Increasingly technology became more *communications-based* rather than *computer-based*. More and more, the value lay not as much in single, stand-alone applications as in the networking of applications, or the communications process itself. The real benefits of office automation, for example, do not lie in providing a single white-collar worker with a desktop computer but in providing a *critical mass* of workers with access to the communications network. The network provides the ability to communicate, retrieve information, and perform all the other familiar office automation functions. Reaching a critical mass is an economics of scope issue. And economics of scope is fundamental to economic behavior. This shift to communications-based benefits underlies each stage along the migration route: computer-integrated manufacturing, office automation, company-wide (enterprise) integration, interorganizational systems, and consumer integration.

Reaching Critical Mass

In the final anlaysis, information technologies alter business in two ways. First, the product or service can be changed and new products or services created. This impact is explored more fully in Chapter 6. Technology's role in changing products and services is a relative newcomer in management's thinking and is not well understood by many people today. The second impact is economic, the primary topic of this chapter. To most people, positive economic impact translates to productivity improvement, and that for many people translates to cost reduction: achieving a given level of output at a lower cost. But the reality is that economic issues of applying technology rarely confine themselves to short-term cost reductions. That is not to say that this use of technology does not have its place. It does. For convenience, we will dub this traditional form of productivity improvement P1. Most acquisitions of computer technology in the past have been justified on the basis of P1 gains.

Productivity is defined as output over input. Thus, the concept of P1 is simple: for a given level of output, input (costs) is reduced. But once one moves beyond P1, measuring productivity becomes an intellectual swamp fraught with all sorts of dangers. In the 1970s, a number of productivity research centers sprang up in response to the realization that the U.S. rate of productivity growth had fallen below that of our major trading partners. As researchers began to probe these issues, they found that productivity is not easy to pin down. At the simplest level, consider the question of whether a 1987 Buick is equal to a 1986 Buick in terms of measuring output. Is a fully equipped Buick equal to a more modest model? Is a Buick LeSabre equal to a Buick Century? How does product quality and customer satisfaction fit into the formula?

These questions probe separate dimensions of productivity. One is that output can change as well as input. Either the quantity or the quality of output could increase for a given level of input. Consider what this means to the economics of technology. If management invests in new technology in the faith that benefits will ultimately be returned in the form of cost (input) reductions for current business, the investment might express itself as an improvement in output (quality or quantity) *without affecting the company's level of input.* We often justify giving managers and professionals desktop computers, for example, so they will be more efficient or effective. But, how does this "benefit" express itself? Perhaps as reductions in cost, perhaps as an increase in output. There is a difference between the two. A manager may think he or she is getting a reduction in costs when the investment actually yields an *unintended improvement in service quality.*

The second dimension of productivity is what happens when a *changing* level of input is combined with a *changing* level of output. This, of course, most closely resembles reality. This dimension comes from the fact that businesses change. They add new products and services, or alter them, or delete them. They operate in times of plenty and prosperity under different constraints than in times of economic hardship. Businesses pass through phases of *innovation* and *experimentation*, taking risks that may or may not pay off, and through phases of tightened control when the operation must be "put into order."

This point is especially important in relation to the assimilation of new technologies. James L. McKenney and F. Warren McFarlan (1982) at Harvard outlined four phases of assimilation. Phase 1 concerns the decision to invest in the technology, Phase 2 concerns experimentation with the technology, Phase 3 is control of the technology, and Phase 4 is the transfer of the technology to other applications. Management response to each phase is critical if stagnation of progress is to be prevented. For example, too much focus on obtaining measurable productivity benefits in Phases 1 and 2 may stifle innovation. Often benefits that accrue from new technology cannot be predicted. Innovation and experimentation are important for flushing these benefits out.

Although it is theoretically possible to segregate different dimensions of productivity and isolate their effects, little practical value can be attached to the exercise. Productivity is a highly dynamic concept which is influenced by changes in customer expectations, the passage of time, competitive forces, and myriad other factors. This type of productivity cannot be subject to precise measure and therefore no purpose attached to the measurement of it. It is system-wide productivity and encompasses the entirety of the company's operations. For the remainder of the book, it will be designated P2.

P1 concerns localized efforts to improve measurable short-term results. As we shall see when we consider the task of structuring a technology portfolio, P1 has its place. But efforts to improve P1 productivity are suboptimal if they replace P2. P2, on the other hand, fundamentally changes the economics of the firm, or at least a portion of the firm. P2 can only be achieved by transforming the entire organization from one way of doing business to another. Thus, the benefits of P2 do not lie in the automation of a single task or job, but in the automation and integration of all tasks.

The obvious question at this juncture is what the purpose of transforming the entire organization would be? And, how would the transformation be accomplished? The purpose is simple enough: transformation must occur if the United States is to compete effectively in the international marketplace. We have already noted that for the first time since World War II, the United States is faced with serious competition. But the real threat comes not so much from new entries onto the competitive field as it does from the nature of the game they bring with them. In addition to invigorating price competition, products from the Pacific Basin and other emerging trading partners have made customers much more quality conscious. And, a number of forces are combining to create rapid turnover in products and services. Recent years have seen dramatic reductions in product life cycles. While many of the inroads from foreign competition have been made in manufacturing, competition is likely to intensify in the service sector as well. One only has to consider, for example, Japan's growing presence in the financial services industry in the United States. Thus, transformation of American businesses must occur in white-collar service sector work as well as manufacturing.

The purpose of the transformation is therefore to alter the capacities of U.S. companies to compete effectively in the international marketplace. This translates into being able to compete economically, matching or exceeding product or service quality, and being flexible in order to respond to the rapidly changing marketplace. The transformation would be accomplished for the most part by building a technological infrastructure appropriate to the company's competitive goals. The infrastructure itself would be defined by the critical mass of technology needed to capture P2 productivity gains—those gains that alter the economies of scale and scope. Management's objective is to define the infrastructure and develop the critical mass. Once in place, then management's objective becomes the *continuous improvement* of the infrastructure and its evolution to meet the changing goals of the business.

Other writers have suggested ideas similar to that of infrastructure. The term "architecture" has become fairly common, for example. Architecture refers to the physical structure of information technologies, clearly recognizing that the structure must be appropriate to the business goals of the organization. As we shall see, there is an intimate relationship between technology architecture and organizational structure. "Information utility" is closely akin to concepts of infrastructure and architecture. Conceptually, information utility shares similarities with water, gas, telephone, and other utilities in that it implies a delivery system and clearly defines the relationship between service provider and receiver. Whereas architecture relates to the physical structure of technology, information utility describes its role. The information utility concept is a powerful precursor of technology's future and provides many clues about how the economics will behave. Many organizations will be restructured (transformed) along the lines of the information utility.

Both architecture and information utility are encompassed by the technological infrastructure. The primary difference is that infrastructure is a more holistic approach. It extends to include other factors besides technology or, for that matter, organization structure. The entire plexus of organization culture, control systems, management style, human relations, and public policy issues is seen as a part of the technology-related decision-making process. These issues will be explored more fully in Phase III. Indeed, the Integrated Technology Management Framework presented in Chapter 2 makes these issues an explicit part of the overall process of managing technology. A key feature of the infrastructure, for example, is how architecture and information utility are used to meet short-term goals of the organization in *addition* to making it competitive over the long haul.

Building the Manufacturing Infrastructure

By far, managers are more familiar with the office than manufacturing, and for that reason it might seem logical to begin with a discussion of office automation. As it turns out, manufacturing is the better place to begin the discussion of building a technological infrastructure, for the reason that the method and benefits are easier for most people to understand. Because manufacturing is physical, its transformation can be more lucidly articulated. As it turns out, building the office technology infrastructure presents the same problems and challenges as those encountered in manufacturing. Conceptually, office automation and computer-integrated manufacturing are one and the same from a management perspective. This means that managers who have no responsibility for manufacturing still have an incentive to understand manufacturing automation because it provides a sort of model for office automation. It is interesting to note that an increasing number of managers refer to their office technology with terms like "data processing factory."

Manufacturing without computer integration depends on large-scale production quantities to achieve economies of scale. Although much of pre-computer manufacturing uses automation, the equipment is physically structured to perform a specific task or is guided in that task by a human laborer. The production line is rigid. In order to change the product being manufactured, for example by going to a new model year for an auto, the line has to be taken down and retooled. Weeks are often required. Flexibility is limited. Another characteristic of pre-computer-integrated manufacturing is that the system is complex and difficult to control. Processing is rarely smooth and interruptions are frequent. Interruptions caused by machine failure, for example, create widespread problems throughout the plant. To offset these interruptions managers seek to buffer operations with inventory. Sufficient raw materials and work-in-process inventories permit sections of the plan to keep doing their job when a feeder operation falters.

Many managers still think of computer-integrated manufacturing automation in terms of replacing human workers with robots. Yet labor costs account for only 5 to 15 percent of total manufacturing costs. Even if labor costs were a larger component, little direct benefit would accrue by simply replacing laborers. The savings is quickly offset by the cost of the robot, installation, programming, and maintenance. The real benefit lies in three areas: (1) continuous production, (2) quality, and (3) flexibility and scope.

Continuous Production

By using computers to coordinate and integrate various functions in the factory, continuous production replaces incremental production. When breakdowns do occur in the computer-integrated factory, the computers simply route incoming materials to another segment of the factory to be handled. In a sense, the computers configure the factory dynamically as the needs change.

What permits the computer to accomplish this configuration? The answer is that all manufacturing equipment in the plant is programmable; the equipment takes its instructions for performing a physical task (i.e., fabrication, assembly, extrusion) from a computer rather than a human operator. This means that the computer can be reprogrammed so that the device can do a different physical task than the one before, within certain constraints. Devices that can be programmed include not only robots which do assembly and machining, but automated materials handling (conveyor) systems, inventory storage equipment, and computer-assisted design (CAD) systems.

The real advantage lies not in being able to change the program in each device, but in *being able to control all devices and their activities from a single computer or computer system*. This is what is meant by computer-integrated manufacturing. Continuous production means that large stocks of new materials and work-in-process do not need to be kept in readiness in case of an operations breakdown in one segment of the plant. Precision control over all aspects of production permits products to be manufactured upon demand and materials inputs to be delivered only as needed. Most readers will know this as just-in-time inventory. Significant savings come through inventory holding and storage costs. Money previously tied up in stock can be employed more productively elsewhere.

The fact that *all materials* in the plant are in continous production can be used as a criterion for success in computer-integrated manufacturing. In the mid 1980s IBM converted its Lexington, Kentucky facility, originally built in 1957, to the new concept. According to one manager at IBM's facility: "Installing the robots is the easiest part; the problem is *integrating* the whole system."

Quality

Quality occupies a place of particular importance in computer-integrated technologies. Not only has international competition provided a reason for renewed interest in product quality, but there is an economic reason as well. Or, as W. Edwards Deming, a gentleman whose influence on Japanese management has been considerable, says: "Paradoxically, through improvement of quality, guided by consumer research, the ultimate result is not only better quality but also lower cost and improvement of competitive position." (1982, p. 181)

At IBM's Lexington plant, the goal has been zero defects in production. Part of the reason has been the commitment to building a quality end product. Here, the technology's contribution was described by one plant manager who said, "Machines routinely make things the way they ought to be made . . . they are consistent." But improving quality in the manufacturing process has actually reduced costs. When IBM began to install computer-integrated manufacturing at Lexington it found that in order to produce quality products and reduce costs in the refurbished plant, quality inputs were necessary. In some cases up to 50 percent of the incoming parts from outside sources were defective and several employees were on payroll whose job it was to determine whether the company could use the "nonconforming" parts. IBM literally had to work with suppliers in some cases to convert them to the quality viewpoint. One IBM manager described quality as the "linch pin" in the computer-integrated manufacturing process. Or, as another put it, "Mistakes are *not* okay."

Striving for zero defects as an economic goal substantially changes management's role with respect to information technologies and the infrastructure in particular. According to Deming and his followers, managements focus should be on improving "constantly and forever" the system of production and service. In technology, that means continuously improving the infrastructure. Marta Mooney describes this approach as follows:

> Management's focus is changed by redefining the manager's functional role. Traditional management defines this role loosely in terms of "planning," "organizing," "leading," and "controlling." The new approach defines the managerial role tightly—and differently. The manager's role is to *improve the process.* A basic tenet of this new approach states "Workers work in the process. Managers work *on* the process to improve it with the workers's help." From a managerial perspective, this distinction is so fundamental that it warrants persistent emphasis—hence the designation *process management.* (1986, pp. 386–387)

The notion of continuous process improvement matches perfectly the need to incorporate new technological developments incrementally into the infrastructure as they occur. It recognizes that the infrastructure, large and complex as it may be, must change and evolve to be consistent with the company's *current* needs. Continuously improving the infrastructure means that its quality and productivity are continuously improved as well. But, productivity is the *system-wide* or P2 variety.

Flexibility and Scope

Computer-integrated manufacturing offers at least one other important advantage which is changing the flexibility and scope of production. The IBM Lexington plant, for example, manufactures microcomputer components and electronic typewriters. In fact, at last count, *fourteen different products* can be produced on the same line and the potential for others exists. Converting the line from one product to another often takes only hours. And economic order lots are substantially smaller. Some manufacturing experts argue that eventually an order lot of one ("one-off" production) will be economically feasible with computer integration.

Flexibility and scope are not unconstrained, however. The universal production plant that can manufacture any product in any quantity does not exist nor is it ever likely to exist. IBM's Lexington plant, for example, was designed to build products that fit into a 22″ × 18″ × 26″ envelope. The plant exploits size as a criterion for manufacture. It can produce typewriters, keyboards, and printers—items that can fit within the envelope. This concept, that unrelated products share certain fundamental characteristics such as size or other structural and manufacturing similarities has renewed interest in "designing products for automation." In such cases the product itself may undergo certain changes to permit a robot, for example, to perform the assembly. Currently, robots do not have the same degree of dexterity as humans and are not as nimble in reaching hard-to-get-at places.

Such criteria as size will open new horizons for restructuring industries. As computer-integrated facilities are constructed to exploit specific product characteristics, other products that share those characteristics will be brought into the fold in order to optimize use of the facility. Heavy capacity utilization will be needed in any event to recover expensive investments in automation and these new product combinations are likely to make for strange bedfellows.

Before leaving the manufacturing infrastructure, two additional points are in order. First, computer-integrated manufacturing is in its infancy within the United States. Only a handful of plants have been built that qualify as truly computer-integrated as opposed to computerized. American industry is faced with a steep learning curve. As experience accumulates, our understanding of how to achieve further benefits will increase dramatically. Specific examples abound. When robots are programmed for example, they must be taken out of service for testing and debugging before they are returned to production. Recent advances will allow us to simulate the robot's actions as a means of testing the program without removing the robot from the workflow. Artificial intelligence will undoubtedly have a major impact on the overall effectiveness of manufacturing automation. We already see applications for "intelligent" vision systems and expert systems. It is this tremendous unexplored capacity that encourages us to conclude that a technology infrastructure can be built to regain for the United States a leading role in international manufacturing.

The second point concerns the rather artificial exercise of talking about computer-integrated *manufacturing* as if it were something separate and apart. Chapter 3 outlined levels of integration beyond the factory. Integration can extend from the factory into internal systems such as accounting, marketing, and strategic planning—the territory of "office automation." Systems can and do extend to suppliers and customers (i.e., they become interorganizational). And increasingly systems are extending to the consumer. Will we reach the time when an individual consumer can enter an order for an automobile and have that order executed and delivered without human intervention? Absolutely. We are many years from that day, but it will come. Just because the manufacturing process is our immediate concern does not mean we should set it up as an "isle of mechanization" unto itself.

Building the Office Infrastructure

Office automation refers to the application of information technologies to white-collar work, especially knowledge work. In most instances, computing is put directly into the hands of the individual or "end user." From a technical vantage point office automation is simply an extension of the distributed data processing concept discussed in Chapter 3 taken to its logical end. However, for purposes of future discussion office automation includes the following five categories:

Conferencing. Telecommunications systems that facilitate human communication, ranging from basic telephone conversations to video conferencing.

Information Transfer. Electronic message systems which use keyboard characters, facsimile images, or voice.

Information Retrieval. Computer-assisted recall of previously stored information, in the form of data, text, graphics, or audio or video input.

Personal Processing. Interactive computer-assisted writing, editing, calculating, and drawing, including applications commonly known as word processing, desktop computing, and interactive graphics.

Activity Management. Systems such as electronic tickler files and automated task-project management, which track, screen, or expedite schedules, tasks, and information.

Almost all managers are concerned with the problems of the automated office and the reasons for this are easily understood: White-collar work is the most lucrative prospect for improving productivity in the U.S. economy for several reasons. First, white collar workers make up the majority of the total workforce. In 1900, they represented less than one-fifth, rose to about one-third in 1940, and by 1980 they were the majority. Knowledge-workers (managers and nonmanagerial professionals) make up approximately two-thirds of the white-collar group. In terms of dollars spent on compensation and internal support, knowledge-workers represent four-fifths of the annual cost to business of white-collar workers; clerical workers make up the remaining one-fifth.

White-collar workers represent not only a big target for automation, but an underdeveloped one as well. Over the last half century, for example, capital expenditures have been much higher *on a per capita basis* in agriculture and manufacturing than in the service sector where most white-collar workers are concentrated. Agricultural workers have been capitalized at over $50,000 each; manufacturing at $30,000; and service at only $2,000 or so. The rate of productivity growth has reflected the level of investments, with agriculture running 70 to 80 percent, manufacturing 35 to 40 percent, and service 3 to 5 percent.

In the United States, automation of white-collar workers may also be important because of American management's penchant for numbers and other information. Following World War II, middle management ranks grew even faster than revenue in most companies. Middle managers, who were originally responsible for translating senior management's goals and policies into operational programs, increasingly became data managers. It was they who collected, analyzed, and interpreted information and passed it on for top management's consumption. Middle management staff often came to dominate in decisions about line operations, and undoubtedly the growth of this middle layer was closely associated with the rise of "sophisticated financial controls" in most large companies. Clearly, middle management has outgrown its purpose and has become bloated. By 1980 managers and administrative staff in the United States accounted for approximately 10 percent of all nonagrarian employment; in West Germany, 3 percent; in Sweden, 2.4 percent; and in Japan, 4.4 percent. All in all there was an increase of more than 50 percent between 1947 and 1980 in the number of nonproduction workers as a percentage of total employment.

With the stakes so high, managers have been particularly anxious to make technology pay off in the office. Expectations are high. In the Fordham survey, for example, over 90 percent of the respondents agreed that information technologies will *significantly* change professional productivity in their company, and three-fourths agreed for managerial productivity. Some companies have indeed had notable success, although these experiences have not been typical. In general, the penetration of the technology especially into the ranks of knowledge-workers (i.e., managers and professionals) has been low. Only about one-third of the support purchased from the information industry for white-collars goes to knowledge-workers, who account for two-thirds of the total workforce by count and eight-tenths by compensation.

Also, as discussed earlier in the chapter, the fact remains that white-collar automation does not seem to be paying off. Most computers, about 90 percent, are used for nonmanufacturing processes. Hundreds of billions of dollars have been spent on automating offices, and, according to William Bowen:

> So far productivity has grown more slowly in the computer age than it did before computers came into wide use. Growth in white-collar productivity has been especially weak, and white-collar work is also where most computers are. (1986, p. 20)

Many reasons for these disappointing results from office automation have been offered. Managers are resistant to the technology. American business is still learning how to use the technology—the "X-ineffeciency" factor, as one expert termed it. The workforce must be trained to the new technology. But these rationalizations still fail to explain why massive investments made since the 1960s have not had a significantly measurable impact. Instead, I propose that we examine two relatively straightforward questions which provide much of the answer: (1) What benefits do we expect from office automation? and (2) What elements of office technology must be in place to generate these benefits?

What benefits do we expect from office automation?

Numerous studies of office automation's potential benefits have been conducted over the years. One of these studies (Poppel, 1982) was conducted by Booz, Allen & Hamilton, the large consulting firm. The Booz, Allen study was remarkable for a couple of reasons. First, a large amount of data was collected on a broad range of issues. Approximately 90,000 time samples of knowledge-worker activities were collected from 300 professionals and managers in fifteen large U.S. companies. Mail surveys were also conducted with several hundred organizations. Second, the Booz, Allen study attempted to quantify the benefits expected from office automation and, based on these results, to make recommendations concerning what actions managers should take as a result. This is important because the study allows us to determine how the benefits are to be derived and then to consider whether office automation technologies can respond in the manner supposed.

From the data collected, the study pieced together a time-profile for knowledge-workers. By far the knowledge-workers spent most of their time (46 percent) in meetings. Another 29 percent went into intellectual work (i.e., reading 8 percent, analyzing 8 percent, document handling 13 percent) and the remaining 25 percent into what the study termed "less productive activities." Less productive activities included totally unproductive time traveling or waiting for a meeting to begin or a machine to become available), quasi-professional activities (seeking information or expediting tasks), and time that was productive only at a clerical level (making copies, filing, making appointments).

The study concluded that office automation could make a substantial difference: within five years 15 percent of knowledge-workers' time could be cut. Half would come from less productive activities and the balance from meetings, analytical tasks, and document handling. For the cases studied, Booz, Allen recommended that companies spend $8,200 per professional within eighteen to twenty-four months. This would result, they estimated, in a post-implementation payback of fourteen to fifteen months on average producing a return on investment of 41 percent.

What elements of office technology must be in place to generate these benefits?

The Booz, Allen study provides a number of examples of how office technology could be used. It cites, for example, the ability to scan publicly available data bases, to communicate using electronic mail and centralized voice messaging, and to access the latest information. But almost all of the examples assume that some type of infrastructure is in place. Communications networks are needed to access data bases and to use electronic mail and voice message systems. The data bases themselves must exist. In addition, for some of the applications—particularly electronic mail and voice messaging systems—all the people one communicates with must be on the system in order to make its use an attractive proposition. Thus, the recommendation to spend $8,200 per professional on new office systems is meaningless unless part of the money is spent on developing the infrastructure or unless the infrastructure is already in place.

Since automating the office is really no different in concept than computer-integrated manufacturing, it makes no more sense to provide a single personal workstation for an office worker than it does to replace a factory worker with a robot. As in manufacturing automation, the key is reaching a critical mass by integrating all activities with computers: system-wide automation. This permits the organization itself to be redesigned in a manner that increases P2 productivity. In other words, office automation is an all-or-nothing proposition when it comes to achieving the most important benefits. When we ask how many U.S. companies have achieved true office automation, the answer is precious few. And when we ask what productivity gains we have realized among white-collar workers since the introduction of computers, the answer is also precious few. The reason is that we have focused our energies on P1 rather than P2 productivity.

Almost line item per line item, office automation, properly implemented, will produce exactly the same benefits as computer-integrated manufacturing: (1) continuous processing, (2) quality, and (3) flexibility and scope. Continuous production improves manufacturing economics primarily because it squeezes excess inventory out of the system. Continuous processing in office automation does precisely the same. But inventory in the office is time. Time is the buffer for inefficiencies in the workplace or "less productive activities." One need go no further than telephone tag for a prime example. Much time could be saved by simply putting everyone on a voice message system. But, while voice messaging might save most employers some time on the job, the impact probably is not deep enough to warrant redesign of the organization. In other words, no benefits will be *realized.*

Quality is also a key benefit. Managers often justify expenditures on office technology because it will "improve the quality of decision making." Fair enough. However, decision making may need to improve as offices are automated because each *remaining* employee becomes more important in the scheme of things. One manager gave this example: A section head has a staff of ten financial analysts. Even if this section head is not too bright, the underlings will probably carry the weight, overcoming the boss's deficiencies. But with a spreadsheet program, the ten analysts now disappear and only the section head remains. There is now much greater pressure for him to be useful and competent at what he does, since others in the organization are forced to rely upon him. Just as the economics of computer-integrated manufacturing demand quality inputs, office automation does as well.

Understanding Knowledge Work

All this means that before the benefits of office automation can be fully realized in the manner suggested by Booz, Allen and others, the technology infrastructure including architecture and information utility components must be in place. However, that is not to say that simply because of the infrastructure, knowledge-work productivity will increase. The reason is that we generally have a poor understanding of the nature of knowledge work. For example, although reams of literature exist that address what managers are and what they do, little agreement about the definition of management exists. There has, of course, always been a tendency to reduce management tasks to steps: setting priorities, allocating resources, and controlling. Based on this formalistic view, some observers have attempted to assess technology's impact on senior managers. Many authors describe managerial decision making as a stepwise process: problem identification, analysis, alternative definition, evaluation, and selection. In this view, the computer, would have a different degree of impact on each activity. The shortcoming of this view is that it fails to recognize the diversity of management; instead, it provides only a one-dimensional view of what it does. Indeed, the strength of any given management team rests in part on its diversity of skills and talents and the appropriateness of this mix to the issues of hand. People become managers because their particular strength—analytical, leadership, interpersonal—is needed.

Over the past few years, some thinkers have begun to recognize that management is not the formalistic process many claim. Kotter concluded that general managers are "less systematic, more informal, less reflective, more frivolous than a student of strategic planning systems, MIS, or organizational design would ever expect." (1982, p. 156) Much of the so-called planning and organizing done by senior managers is hit or miss, he concludes, done is a rather sloppy manner, and thus a lot of managers' behavior can be classified as unproductive.

But—Kotter believes that hit or miss is precisely how planning and organizing are carried out by executives. First, they must figure out what to do in the face of uncertainty, and then try to get things done through a large, diverse group of contacts. Initially, in a new situation, the manager will develop an agenda, then build a network of cooperative relationships, and then induce the network to execute the agenda.

We have also long recognized that collecting "soft" information is another valuable activity performed by management. Such information is likely to be collected verbally and may be put to little immediate or specific use. Much is collected during casual, even chance, meetings and conversations and in the context of activities that the Booz, Allen study and others characterize as "unproductive."

Finally, what managers do with information varies widely. Evidence suggests that managers differ in the way they gather and evaluate information. For example, prospective gatherers tend to look for certain elements in the communication. Receptive gatherers tend to just listen, later combing the information for meaning. The method for evaluating information can be either systematic or intuitive. A systematic person has a predefined method of analysis, whereas the intuitive evaluator approaches each communication differently.

Two important conclusions can be drawn from these viewpoints. First, knowledge-work activities categorized as "unproductive" may in fact be productive. No one would argue that some activities performed by managers and professionals are clearly unproductive, and that their productivity could be improved. Perhaps the clearest example is when support services are inadequate and a manager must stand in line for a turn at an overutilized copying machine. But such obvious inefficiencies are the exception. They are not likely to be the basis for significant productivity improvements nor justification for large expenditures on information technologies.

However, the many so-called unproductive activities actually provide the context for important—even critical—exchanges of information, hard or soft. They may also be important contributions to the networking activities needed to implement an agenda. If this is so—if we proceed on the simplistic notion that a white-collar job can be disassembled, the unproductive elements stripped away leaving the productive tasks—then we run a real danger: the unacknowledged value of so-called unproductive job elements is also being stripped away. In other words, the individual's effectiveness in performing valuable tasks such as exchanging soft information and network building may be dampened.

Even if there were no outward impact such as that described above, it is still not clear that knowledge workers could significantly improve their efficiency and effectiveness. This thinking falls into the foolish trap of attempting to judge knowledge work as if it were piecework. The number of autos produced, items sold, transactions processed, and similar measures may be applicable to blue-collar and clerical employees but such measures are generally inapplicable to knowledge-workers.

In many cases, it is enough for knowledge-workers, especially in senior management, to be. This notion of "management by being" expresses itself in numerous ways. A manager's role includes leadership. His or her presence at a stockholders' meeting, civic gathering, or employee retirement luncheon may be enough inspiration to the function.

Another example of productivity that cannot easily be quantified is work performed by a technology gatekeeper. This is an individual who surveys the environment for technical innovations by attending professional meetings, talking with colleagues and experts, and similar activities. Valuable information, much of it with competitive implications, may be exchanged informally only after a relationship has developed between parties. The time spent getting to meetings or lingering over the water cooler, which the automation enthusiasts deem unproductive, may not be so unproductive after all. Activities we regard as barriers to productivity may not be barriers at all.

The second conclusion is that knowledge-workers are a diverse and complex group. Little is understood about the nature of their work and, more importantly, how technology will ultimately impact it.

Predictions about the impact of office automation have also relied to a great extent on statistics that purport to represent the "average" knowledge-worker. The average knowledge-worker spends 46 percent of his or her time on meetings, 29 percent on intellective activities, and so forth. By averaging these figures across all workers and, worse, all categories of knowledge-workers (i.e., managers, administrators, professionals, sales staff), knowledge work is characterized by a broad set of generalizations. For all the words written on the subject, we do not have a clear understanding of what knowledge-workers, especially managers, really do. This is perhaps, why, managers can be so extraordinarily well rewarded. Beyond that, we have even less understanding of the impact of technology on the knowledge-worker.

The Pipeline Theory of Knowledge

Information technology will affect the economics of knowldge work in another way: by automating *knowledge* itself. An important distinction exists between information flows and knowledge within an organization. Most of the gains we have focused on thus far with regard to the integration of both factory and office derive from the automation of information flows. Previous gains from the automation of transactions processing fall into the same category. The types of activities addressed by these automation efforts were largely routine and structured; tasks typically handled by clerical workers. Indeed, some people believed that computers had little in the way of a contribution to make beyond this role. John Dearden, writing in the 1966 *Harvard Business Review,* commented:

> It is my personal opinion that, of all the ridiculous things that have been foisted on the long suffering executives in the name of science and progress, the real-time management information system is the silliest. (1966, p. 123)

The long-suffering executive not withstanding, designers of information technologies have increasingly focused their efforts on knowledge work. Management information systems (MIS) were developed to report on the firm's internal operations and business environment. Decision support systems (DSS) attempted to address relatively unstructured decisions. But MIS and DSS more often than not accomplished their purpose by summarizing and quantitatively manipulating transactions data. Office automation, of course, addresses white-collar (including knowledge) work but concerns support activities (e.g., communications, document processing, activities management) more than the content of knowledge work itself. But now, with the advent of expert systems technology, much of that promises to change.

Knowledge is the "know-how" in the organization: the fact or condition of knowing something with familiarity gained through experience or association. An organization's base of knowledge often distinguishes it competitively from others. We have already seen in Chapter 4 that many companies resist the transfer of knowledge because they fear their ability to compete will be compromised. Organizational knowledge may consist of expertise in design or production of a product, marketing know-how, strategic prowess, financial ability, or any combination of these.

Most managers equate organizational knowledge with human resource issues primarily because expertise typically resides with individuals or groups of people. We therefore speak of bringing "fresh blood" into the organization, meaning people with new ways of thinking. We speak of getting rid of the "dead wood," meaning people whose *usable* knowledge is marginal at best.

Indeed, given that people are the carriers, knowledge flows fairly freely into the organization and out again when individuals are hired and when they leave, respectively. One might envision this process as an open pipeline (Figure 5-1a). The pipeline represents the level of knowledge within the organization at any given time. At the back end of the pipeline, knowledge flows out as people retire, resign, or die. Knowledge is input either because the company hires experienced people away from other concerns or because the company hires new recruits—the "green" college graduates. Either method of obtaining knowledge is expensive. The new hire must be indoctrinated into the organization and his or her expertise developed to and maintained at a useful level. If the stream of incoming talent is stopped, as in hiring freezes during economic downturns, the organization's knowledge will atrophy with time. Likewise, if an excessive

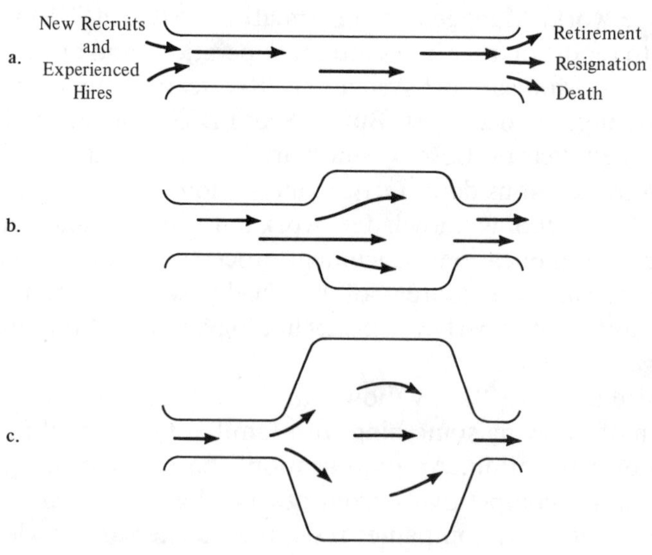

Figure 5-1. The Knowledge Pipeline.

number of terminations occur and too much expertise leaves too quickly without being replenished, knowledge will also atrophy. As an example of the latter case, companies have sometimes offered early retirement packages only to see their best and brightest employees leave. Thus, over time some parity must be reached between the inflow and outflow of knowledge.

Certainly, any technology which could capture human expertise in a useful form could radically alter the economics of knowledge work. Essentially, by transferring knowledge from individuals to another vehicle, one under the control of the organization, its outflow through employee retirement, resignation, and death would be stemmed, as if one put a dam or filter within the pipeline to control the flow. In Chapter 3, we discussed a technology capable of capturing knowledge: expert systems, a branch of artificial intelligence.

Three examples of how *expert systems* capture human expertise will suffice:

- General Electric (Fairfield, Connecticut) built a repair consultation system known as DELTA or CATS-1. The system identifies causes of steam locomotive malfunctions and gives repair advice.
- Schlumberger (New York) developed DIPMETER ADVISOR a system which interprets geological oil field data.
- Apex, a joint venture between the Travelers Companies and Integrated Resources, created a system called PROPLAN which provides personal financial consultations.

Numerous other development efforts are proving daily that expert systems technology can be applied successfully to any area of human expertise. As this technology's influence spreads over the next couple of decades, many knowledge-workers' jobs will be automated as clerical jobs were in the 1960s and 1970s and, for reasons outlined in Chapter 3, further developments in the technology will ensure the momentum. As the technology takes effect, knowledge will be increasingly retained within the organization (Figure 5–1b). Gradually, fewer and fewer people will be needed simply to *maintain* a given level of knowledge. Fewer new employees will be added and eventually the number terminating will dwindle as well (Figure 5–1c). In other words, in order to maintain a given level of knowledge the human expertise required will decline. However, for the same level of human expertise (no change in the knowledge-worker level), knowledge will increase. Either way, the economic impact alters.

The impact may be felt in a different way. As less and less human input is needed to *maintain* a given level of knowledge, the focus of knowledge work activity will become *leveraging* expertise. In other words, the task of the human will be to expand knowledge. As knowledge or expertise is expanded, expert systems technology should capture it. This process is likely to be aided by developments in the technology itself as expert systems learn how to improve their problem-solving capabilities through their own initiative.

CONFLICT BETWEEN ECONOMIC BEHAVIOR AND CURRENT MEASUREMENT STANDARDS

The shift to communications-based technology has created a second problem, since reaching the point of critical mass takes time. If it is true that the most lucrative benefits of office and factory automation flow only once a critical mass is reached, then the real pay-off from communications-based technologies may take longer than earlier technologies. In other words, with communications-based technology, obtaining benefits may be an all-or-nothing proposition. Little hard evidence is available to support this conclusion, but two arguments can be advanced on its behalf. First, already cited is evidence that office technology is not increasing white-collar productivity as much as we had anticipated. Few companies have achieved the truly automated office which is fully integrated using communications technology. Thus, many of the applications such as electronic mail, filing, and calendaring which would generate large productivity gains are not yet widespread. And, since the preponderance of white-collar work is communications—almost half—unless features of office automation that affect communications are in force, gains are unlikely.

A second argument is that there are some very convincing reasons why reaching a critical mass in factory or office automation would produce additional benefits. At that point, the technology would allow for a complete reorganization of the company or unit to occur. In other words, the benefit comes not from piecemeal application but from the ability to design the entire work process differently. The effect might be summed up by the following question: "If I started the company today, given the current capabilities of information technologies, how would I put the organization together?"

Unfortunately, it may take a long time to bring the information technology system to a critical mass. And time is the enemy of the American manager if it means giving up profit in the interim. All things being equal, major investments in technology that delay return on investment have an adverse impact on the company's reported performance. The emphasis on short-term performance is, of course, legendary. Net income reported in its various forms is still the most important measure of financial performance, although cash flow is a close second. Accounting reporting standards reinforce the prominence of income through measures such as earnings per share and these figures play a significant role in the investment community's evaluation of corporate securities. Quarterly reporting requirements ensure that managers in affected companies plan on 90- to 120-day cycles.

What all this means is that the pressure for short-term results may be chafing against the need for communications-based technologies that will yield certain benefits only after long-term investment. Or, as one of the Fordham survey respondents commented:

> The investment community's insistence on quarterly and annual bottom-line improvements often cause long-term technology investments to be ignored. Funding of technology is viewed as a cost rather than an investment. Rather than treating the investment in technology as a long-term necessity, it is viewed in terms of a short-term expense which must be controlled. This short-term costs-benefit approach stifles experimentation and innovation.

In Japan, long-term investments in technology are the rule. Japanese businesses use three criteria for judging success: market share, growth, and total sales. Measures of income, like return on investment, are really not used except as a way of ranking internal investment opportunities. Japanese companies will sometimes forego profit in order to capture market share. There are in fact disincentives for Japanese businesses to report income: tax rates run 50 to 60 percent. Japanese firms are typically leveraged more than American companies. They may have a debt to equity ratio that would spell financial disaster for an American firm. But these high debt ratios are understandable when one realizes that the cost of capital is roughly half that in the United States. In Japan, banks are often active owners of major businesses which are organized into families with extensive mutual stock ownership. Seventy-five percent of the stock is held this way. The Japanese investment community reinforces the system. Individuals within Japan are typically not active investors. Even pension funds, which represent large blocks of investment capital in the United States, are not funded separately. Nor is there much choice between income and growth investment opportunities as would be the case for the U.S. investor. Income, from the Japanese perspective, is simply not an option. All investment is for growth.

Perhaps the Japanese businessman who was sharing tea with my American friend could afford to laugh. He has great latitude to experiment and innovate with technology because the investment environment is forgiving. He has the luxury of following his instincts. American managers, despite the lip service given to freedom of action, have that option only if they are willing to take risks with reported earnings. Some are taking that risk, of course. But the point is, why do we require them to take this risk?

This question is the heart of the issue in managing technology. My argument is that an American manager is *handicapped* in competition with foreign companies from countries like Japan. No matter what decision he or she makes with technology, it will almost always be wrong unless it defies conventions for success mandated by American financial accounting standards. Contemporary technology systems take time to put together and bring to operational condition. This is especially true of communications-based systems. Japanese can abide time. Americans cannot. Contemporary technology systems do not produce real benefits until they reach critical mass. Japanese can forego benefits for a period. Americans cannot. Contemporary technology systems cost money; they are capital intensive. The American cost of capital is expensive. In Japan, it is not.

The net result is that all signals in the current accounting measurement system are scrambled; they give signals contrary to the economic realities of the technology. And the manager who recognizes those realities and obeys them must risk the bottom line.

But how could accounting and financial performance standards be so out of tune with the economic realities of technology? The answer requires a brief digression into the history of accounting measurement itself. As we shall shortly see, measurement standards took on a life of their own. In their current status quo, no matter how deeply ingrained in American business thinking they are, these accounting standards are destructive to the American economy and cause American managers to be dysfunctional in decision making. Every American manager knows full well that corporate decisions are often made purely on the basis of their income statement and balance sheet impact. And, the results are not always unwelcome. Managers have been able to smooth income, produce bogus profits in the short-term, and perform other bookkeeping tricks. Thus, while many managers have complained that reporting requirements are onerous, an articulate constituency for reforming accounting measurement has never materialized.

Managers have come to see accounting measures as either a means to play numbers games or an irritation, but rarely as critical to economic success. Now, that is changing. Current standards are helping to erode our competitiveness by causing managers to make wrong decisions about technology. It therefore becomes imperative that managers begin to view accounting and financial standards not as inviolate measures of corporate success but as fallible rules now grown hopelessly out of date. How did these measures come to be out of date?

A Brief History of Accounting Measurement

The Industrial Revolution in the late 1800s brought with it the rise of the large corporation. But even in the early days of large corporations very little measurement of capital asset performance took place. Prime costs—direct material and labor—were the chief concern. These early days were still dominated by individual ownership of firms engaged in single product lines. Financing was largely from internally generated funds. Consequently, according to Robert S. Kaplan, "The effect of the new investment on reducing prime costs or in improving the operating ratio was deemed sufficient to guide the investment decision." (1984, p. 393) Operating ratio was calculated as an operation's costs as a percentage of sales with no allocation of capital costs to products or periods.

By the early 1900s corporations were becoming large, diverse, geographically dispersed entities. The concept of a modern management structure was born, and the management function began to separate from ownership. It was during this period that many current performance accounting practices were born to deal with the growing complexities. According to Kaplan, the appearance of vertically integrated, multiactivity firms in the early 1900s created the need for modern managerial control practices. "Scientific management" advocates began the practices of measuring and allocating overhead costs to products primarily in an effort to develop standard unit costs to implement their methods. By 1925, many of the cost concepts used today for performance measure were clearly established: cost behavior, relevant time period, sunk costs, incremental or differential costs, break-even analysis, and so forth. But, most importantly, thinkers of this era clearly recognized that internal accounting measures served a different purpose from financial accounting, which addressed reporting to outsiders (i.e., stockholders, creditors).

It was also during this period that return on investment (ROI) was born. The DuPont Company devised this measure as a means of evaluating divisional performance in their decentralized, functional organization. Over the ensuing years, DuPont and General Motors refined and elaborated the measure and, by 1923, the basic ROI concept as we know it today was firmly in place. ROI as a divisional measure is the basis of much of management's behavior. It is the basis on which sales and operating budgets are developed and managers allocate resource costs and devise elaborate transfer mechanisms.

As long as this system reflected the underlying economic realities it represented, it contributed more good than harm. But even during the early days, some people cautioned that allocated costs could run from 100 to 125 percent of direct labor, and that distortions were large if these allocations were inaccurate. Even then, direct labor often represented less than 20 percent of total manufacturing costs and thus allocating on the basis of direct labor was a dicey proposition at best. As we shall see, with computer-integrated manufacturing methods the ratio of capital or fixed costs to labor goes up relative to previous methods of manufacture and this puts the notion of allocating costs in this manner further out of acceptable bounds. In other words, these allocation schemes are less and less reflective of modern economic realities.

Another problem relates to the use of the book value of investments based on historical costs to calculate ROI. Older assets, those which have accumulated considerable depreciation, yield higher returns, everything else being equal. Newer assets which have greater book value due to low depreciation depress ROI. In this sense, ROI contains a systematic bias against new investment. Some companies have attempted to resolve the problem by using current asset values or undepreciated costs.

As a free-floating technique to assist management in decision making, ROI has on the whole been a useful measure, given that income is an important objective in corporate performance. The changing mix of labor, material, and overhead costs and behavior would ordinarily have presented management with little problem except the need to revise the method of calculation from time to time.

And these measures would likely have led a pretty much independent existence in the organization, being reserved for internal decision-making purposes. If allowed to evolve freely, a contemporary version of ROI and similar measures might be available which would give adequate, if not perfect, guidance. But they were not allowed to evolve freely; instead they became institutionalized as a part of financial accounting standards following the 1930s economic crisis. These measures were frozen in form relative to economic realities before World War II. After World War II, the economic realities changed.

Since the appearance of ROI, few developments in accounting performance measurements have occurred that have had a significant impact on practice. The only notable exception is the discounted cash flow (DCF) technique introduced in the early 1950s. DCF has become a widespread practice in many corporations, replacing payback and similar methods. The process of discounting future streams of benefits and costs to their present value is referred to by many as cost-benefit analysis. Cost-benefit analysis has been the primary vehicle for justifying investments in computer technology for the past three decades or so.

As a criterion for evaluating investments, cost-benefit has serious drawbacks. Foremost is the fact that it biases the evaluation in favor of near-term net cash inflows. The farther a benefit is in the future, the more heavily that benefit is discounted. Indeed, the basic assumption of the technique is that near-term cash inflows have a greater "time value" by definition. A second problem is that benefits in the distant future (five to ten years hence) tend to be ignored during applications of the technique. Firms usually use the technique for time horizons of five years or less. As we shall see, these two shortcomings, which combine forces to emphasize short-term rewards, may all but exclude any realistic consideration of long-term investments needed for creating true office automation or computer-integrated manufacturing systems.

The reason is that incremental investments may be needed for five years or more to reach a critical mass in these technologies. Only when the critical mass is reached will these communications-based technologies yield up their most important benefits and thus make the investment worthwhile. Time is needed to reach critical mass for several reasons. First, the entire system, including organizational culture and structure, must also be changed and this is a tedious, time-consuming process. Second, large system-wide integrated technologies are difficult to design and implement. Often considerable time is required to work out the kinks and get the system functioning smoothly.

The simplest of examples will suffice. Two years ago, I moved into a new apartment tower in Manhattan containing forty-two residential floors served by four high-speed passenger elevators. The four elevators constitute a single system and at a basic level, the problems of getting them to work together are no different from integrating a manufacturing or office automation system. It took almost a full year before the elevators were functioning smoothly. Engineers were forever tinkering with and adjusting them. In the third month following the building's inauguration, residents were given a 20 percent reduction in rent to compensate for the inconvenience. Many residents were waiting for a half hour or more for elevators to arrive. Today the system works acceptably. Comparing four elevators to a computer-integrated office or factory with hundreds and thousands of electronic devices creates some sense of perspective about the magnitude of the problems involved in making complex systems operational and realizing the benefits.

For the record, management accounting thought has developed in other ways. Examples include residual income, quantitative models, information economics, and agency theory. Information economics differs considerably from economics of technology issues in that the former is a theoretical approach to selecting an information system to aid decision making in an uncertain environment. But since the 1920s academics have been largely responsible for developments in the field and few have had any real impact on practice.

In summary, most economic evaluation of technology in the United States is based on return on investment and discounted cash flows. Both techniques have systematic biases which stress underinvestment in new technologies that enhance the overall capacity of the company's operations or which preclude investments that require long-term commitment. But this is only a fraction of the problem; other historical developments in accounting measurement have a bearing on our current economic perspective.

If return on investment and discounted cash flow are defective measures of true economic behavior for new communications-based technologies, then one might have faith that American managers would recognize the problem for themselves and act to correct it. But faith dies when one realizes that the assumptions that underlie these techniques—the primacy of earning income and short-term performance—have been cast in stone by modern financial accounting reporting whose steward is, of course, the accounting profession. Far from being the product of reasoned human thought, the accounting emphasis is a freak of history. It was born at a critical juncture in the development of the modern, diversely owned corporation and diverted the system's energy into an unproductive direction. By focusing on income, the system assures that productive vitality will be slowly but inexorably bled away.

Before the rise of the modern corporation in the first part of this century, business enterprises existed to produce income for their owners. Previously, wealth was associated with owning property, and indeed biases against entrepreneurs existed among landed gentry. The Industrial Revolution created a new class whose wealth derived from ownership of large commercial enterprises which increasingly were incorporated. These early corporate leviathans remained the property of a single owner or small groups of investors. For them, earning current income was an important motive. In the 1910s and 1920s, corporations began to attract increasingly large groups of investors. This development gave rise to significant abuses. Holding companies and investment trusts were created where funds were overextended or the enterprise undercapitalized. Paid-in capital was used to pay "dividends" to make deals attractive to prospective investors even when no income had been earned.

At precisely the same time that American corporations began their transition from concentrated ownership by a few to widespread ownership by many, the process was seized by speculation. This speculation was possible in part because uniform accounting standards or even an agreement that uniform accounting standards were desirable was largely absent. Information about corporate accounts was often kept private, since audits by independent parties were not required. Emphasis was on the balance sheet. Both balance sheets and income statements were inadequate and arbitrary measures, and as a result distinctions between capital and income were often blurred. The tide of speculation which began in 1924 rose to a frenzy until October 24, 1929—the day commonly known as "Black Thursday" when the stock market failed. The Great Depression followed.

The ensuing economic disaster naturally gave rise to a backlash in which members of the financial community were blamed. The Securities Act of 1933 and Securities Exchange Act of 1934 began to create the accountability framework necessary for regulating public trading. One of the provisions of the 1934 Act was the creation of the Securities and Exchange Commission, which was given sweeping financial reporting authorities. According to Robert Chatov, who has written an excellent history of corporate financial reporting, the newly-minted commission delegated this responsibility to the accounting profession. The full initiative for developing reporting rules was passed to the accountants, the very people the rules were supposed to regulate. In this manner accounting came to be known as a "self-regulating" profession.

The first major task of the newly self-regulating profession was to develop written standards to consolidate their authority. In the absence of a more promising approach, accountants began to record what was then considered to be common practice; the results came to be known as "generally accepted accounting principles." At first, the profession had difficulty convincing all its members and the financial community at large that they should be constrained by such standards. But over the years, a number of events and pressures moved the accounting profession away from the relative freedom of action it had prior to the Great Depression.

Increasingly, accountants were held responsible for the consequences of their professional activities. One notable example was the McKesson & Robbins case in which one of the largest accounting firms, Price Waterhouse & Co., was found to be negligent in discharging its auditing responsibilities. Increasingly, pressures from this 1939 lawsuit and other forces caused the profession to standardize and institutionalize accounting methodology. Through a succession of regulation bodies—the Committee on Accounting Practice, the Accounting Principles Board, the Financial Accounting Standards Board—it created an elaborate structure of principles, rules, and regulations to guide and insulate the profession. And in the process, earnings were raised to the status of chief criterion of corporate success. According to Chatov:

> The importance of earnings became institutionalized. In the future, varying accounting devices would be invented or adapted to inflate or to give an appearance of greater corporate earnings. During the conglomerate merger movement, this widespread practice was to focus attention once again on the accounting scene. (1975, p. 142)

In recent decades, the burden of accounting standards on U.S. businesses has become enormous. With ever complex requirements like earnings per share and interim reporting, the focus on short-term performance has become a vice grip. Any American company that makes a sizable commitment to building up its technology base must do so at the expense of earnings, and a drop in earnings brings criticism of asset underutilization and—in dire circumstances—takeover actions. Supporters of income measures argue that investors are "sophisticated" and recognize when management makes a worthy commitment to long-term investment. But little evidence exists to support that view. And even if that view were accurate why should accounting measures not reflect the value of these long-term decisions? If investors value long-turn viability, why emphasize short-term profitability?

The pressure to generate income has clearly caused American businesses to deemphasize research and development and reinvestment in capital assets, and in a real sense to borrow from the future. In historical perspective we can clearly observe how this subversion of the economic system began to take its toll in the 1960s. After income-oriented measures took root in the 1930s, we might have seen the effects in the late 1940s or early 1950s had it not been for World War II. But the war caused a massive infusion of innovation and technology which provided the industrial base with a shot of adrenaline. This charge was felt throughout the 1950s and into the 1960s. By then the rate of productivity growth in the United States had begun to decline, and since then it has not returned to healthy levels.

Somewhere along the historical trail—with all the confusion of the investment scandals, the Depression, World War II, and a dazzling post-war economic boom—we missed a turn. When corporate ownership began to diversify and grow, the need for income became less relevant. That form of ownership makes most sense if maximizing the wealth of the organization is the primary goal. Remember that the Japanese place little importance on profit per se but emphasize market share, total sales, and growth. The proof of the assertion is beginning to show itself in the concerns of institutional investment managers, those responsible for large pension and similar types of funds. Institutional investors have come to dominate the markets in recent years. Prior to that most investments remained in the hands of individuals who may have had some interest in income. But the sheer size of institutional investments prevents rapid trading of stocks and other instruments. The amounts involved are simply too large to move in and out of particular investments quickly. Thus, the gains must be realized in terms of the long-term performance. In other words, some evidence exists that short-term income per se is becoming less important.

What does all this mean for managing information technologies? It means that in order to achieve the major benefits of technologies like office automation and computer-integrated manufacturing, managers must make long-term investments, not simply major investments, because evidence shows that big, complex technology systems take time to put into place and bring to full operational performance. But managers who undertake this strategy must run the gauntlet of financial community criticism as Roger Smith, Chief Executive Officer of General Motors, has recently done. They must bear earnings cuts and weather criticisms from investors and others. And they must run the risk of making their organizations vulnerable to takeovers and other unproductive financial attacks.

Without exaggeration, the imposition of short-term interests on American businesses is currently the single, biggest constraint on effectively managing technology. Until this constraint is removed, it will be exceedingly difficult for managers to make decisions consistent with the realities of technology's economic behavior and which will allow the United States to hold its own in international competition. And, frankly, because the short-term income orientation is so deeply ingrained in American business thinking and institutionalized in accounting and investment practices, change will come only with Herculean effort. Even then it will be excruciatingly slow. Although there have been some stirrings in the accounting profession in recent years—first admissions of culpability—little has been or will be done to improve the situation in the near future. The question, therefore, becomes where that leaves the technology manager today.

Managers are left with three mandates. First, managers must come to understand the fundamental economic behavior of communications-based technologies such as office automation and computer-integrted manufacturing. Chapter 3 laid the groundwork for this understanding by describing the migration of information technologies and how they alter business practices through automation, disintermediation, and integration. Many of the benefits that derive from this triad come only once a critical mass of technology has been reached in the organization. The reasons for this will be explored in more detail shortly.

A second mandate is that managers must educate the American public in general and the investment community specifically about the trade-offs necessary to sustain long-term profitability using technology. The ability of American managers to do this depends directly on the success of the first mandate: understanding technology's economic behavior themselves. So critical is this step to the long-range success of technology management in the United States that Chapter 9 has been devoted to a fuller exploration of the context in which management must work—the forces that shape the business environment of the company.

A third mandate is that managers must carefully construct an investment portfolio in technology that strikes the best possible balance between conflicting demands for short-term results and for long-term competitive viability. Just as an investment portfolio of financial securities must begin with a goal (e.g., aggressive growth, conservative income and growth, etc.), and be structured with the proper mix of gain potential and risk, so must the technology portfolio follow similar criteria. Given the goal of the portfolio, managers must attempt to maximize performance, whether they are managing financial instruments or technology.

6

Investing in Technology

Technology's Evolving Role in Competition

Over the past few years, interest in information technologies as a competitive tool has moved front and center. It is clear that American business has pinned its hopes on technology and expects great things from it. In the Fordham survey, for example, 85 percent of executives who responded said that their companies would make *major* investments in new technology in the future and almost as many (80 percent) explicitly stated that information technologies are important to their competitive position. Very few respondents would agree that their business had been so successful in the past or would be in the future that they did not need to employ technology.

Productivity improvement is still seen as the primary value of information technologies. Over three-fourths of the managers said computers will be most important as a means of improving internal efficiency and effectiveness. And despite any convincing evidence that technology has had a material effect on white-collar productivity in the past, managers have great expectations for improvements in the future. Over 90 percent believe that information technology will significantly change professional productivity and 75 percent said the same for managerial productivity.

But managers have other expectations as well. Almost 90 percent of the respondents said their companies would use information technologies as a means of strengthening relationships with customers. Two-thirds said the same about supplier relationships. Over one-half said that technology will *completely* change the structure of the industry. And, over half said the product being delivered to their customers has changed as a result of technology. Half also said they are using it to maintain their position as a cost leader in the industry and 43 percent are using technology to create a market niche.

Many managers believe that technology should be moved forward at all costs. In only 13 percent of the cases did Fordham respondents agree that they had withheld major investments in new technology because the outcome was considered too risky. Fewer respondents (9 percent) had delayed investments because of an uncertain or undesirable impact on employees. About the same number (12 percent) had withheld major investments because people with sufficient qualifications to manage the technology were in short supply. In effect, there appears to be little resistance to forward momentum in management's enthusiasm for technology. According to Michael Porter, "Technological change tends to be viewed as valuable for its own sake, any technological modification a firm can pioneer is believed to be good." (1985, p. 164)

Yet some evidence suggests that managers may be engaging in wishful thinking. At the same time that the benefits were so broadly extolled, Fordham survey respondents were also complaining that their greatest source of frustration with information technology was lack of management involvement and understanding. Several respondents decried management's technical illiteracy and their unwillingness to become literate. Some indicated that management failed to commit themselves to a continual assessment of technology's rapidly evolving role. In other words, corporate management touts technology as a primary asset, but their actions do not support their words. The problem, in one respondent's words, is the "reluctance of senior managers who offer only lip service to the need for technology and do not really take the time to understand it." According to another respondent, there is a need to help "senior management visualize the potential changes that could impact the environment and industry over the next five years."

Evidence suggests that American managers still lack any real maturity in using information technologies as a competitive tool. Indeed, the whole notion that technology is a competitive force, more than a means of improving productivity, never really began to take root in management's thinking until the 1980s. Despite the rhetoric about creating "competitive advantage," the truth is that many examples of the strategic uses of technology are indeed no more than 20/20 hindsight. They were less the product of insightful planning and more the by-product of routine applications development. Few started with a well-developed long-range strategy which identified competitive objectives of technology and then proceeded to implement the same. This does not necessarily mean that decisions were bad and that the approach was wrong. But it does suggest that management falls short of achieving much of technology's potential performance simply because they do not know the potential exists.

Indeed, many of the strategic uses of computing have grown out of long-term past experiences with technology which were generally positive and reinforcing. During each stage of technological development, a company may have successfully implemented systems and achieved real benefits. With luck, this process was overseen by truly competent and visionary technology managers who could effectively communicate with business managers. The overall effect of these positive experiences is cumulative so that managers of the technology are permitted to experiment more, taking greater risks in their attempts to apply technology to the business. A level of trust in the technology accumulates. Management is more willing to rely upon technology resources to accomplish critical business objectives. In this environment, technology may become a pervasive force within the company and may ultimately transform the nature of its business as well as the way in which it does business.

The Travelers Companies in Hartford is one example. Travelers is currently becoming a fully integrated financial services concern, but its history is firmly rooted in the insurance business. Like all companies of its ilk, it handles massive volumes of routine transactions generated by sales, policy writing, and claims processing. Or, as one Travelers executive put it; "We live with data." When large data processing operations began to appear in the 1960s, Travelers—like most other insurance companies—substantially reduced clerical costs through automation. Travelers also made a successful transition to interactive processing by exploiting data base and telecommunications technologies. Increasingly the company was able to better profile its customer base, permitting the company to identify new insurance and investment opportunities. Communications networks were put into place to provide better service to its independent agents. It comes as little surprise, then, that Travelers has moved aggressively in attempting to automate its substantial knowledge workforce. Although the process of getting business units to adopt technology has not always gone as smoothly as desired, the company's efforts are nonetheless marked by considerable forward momentum. Yet even some managers admit that they have not moved as fast as they would have liked in realizing the competitive potential of the technology. "We did not realize," one said, "how important giving terminals to agents was in a marketing sense."

American Hospital Supply is another company that gained significant competitive advantage with technology by placing terminals in the facilities of major customers. This was initially done to improve efficiency, but American Hospital Supply's sales soon began to rise. Customers found the terminals convenient and easy to operate and therefore used them more. Another often-cited example of how information technology was used to leverage sales is the airline reservation system. American and United gained significant advantage because they created reservations systems for use by travel agents that listed their own flights at the top of the video screen. Agents who had no particular reason to choose one airline over another often selected flights from the first on the list. Yet this advantage was more a by-product of the systems than the reason for their implementation.

What begins to emerge from a look at these experiences is the realization that information technologies often have an important unintended and unplanned-for impact on competitiveness even though the initial view of the application in question was to improve productivity. So powerful were some of these experiences that managers also began to see that productivity improvement, while it is important, may only be a secondary benefit. Companies like Travelers began to understand and see that decisions about technology were intimately related to their overall business plans. Years of rapid technological change and deregulation of the financial services industries changed the nature of business at Travelers. The conservative tradition-oriented insurance company began to evolve into an integrated financial services company offering a broad range of products and availing itself of numerous new investment options.

The pace of change accelerated and financial service products became more responsive to the consumer market.

Current Thinking on Strategy and Competitive Advantages

Traditional thought on strategic planning is based on an "ends-ways-means" approach. According to this view, corporate objectives are set (ends), strategies developed (ways), and resources marshaled (means). The whole plan is translated into programs, and these have often formed the basis for automation projects. Many businesses do, of course, attempt to use this approach but with mixed success. And critics, such as Robert H. Hayes of the Harvard Business School, argue that it has its limitations and that it does not fully reflect the realities of how a business captures the lead over its competitors.

For periods of less than five years into the future, the goals that can be achieved by this approach are not likely to produce a sustainable competitive advantage. Time horizons longer than that require "strategic leap" mentalities within the organization. Large expenditures are committed to producing major breakthroughs in adapting a new manufacturing technology, lowering costs, or whatever. Yet breakthroughs do not always occur when needed and the result is unmet expectations.

One example of the "strategic leap" mentality is General Motors' plan to leapfrog the industry into a cost competitive position on small automobile manufacture by building the Saturn plant in Tennessee. Saturn not only relies on building new facilities and training a workforce unfamiliar with heavy manufacture but literally creating a new, complex, and largely untested technological production system as well. General Motors has found since beginning the project that its expectations must be modified and that the process will have to be built up much more incrementally than originally thought. Some have also questioned whether the plant—which will not be operational for years hence—will allow it to meet the competition faced by the company.

International competitors such as Japan and Germany and an increasing number of U.S. firms have taken a somewhat different approach to strategy that emphasizes incremental and constant improvement over time. According to Hayes (1986) this essentially reverses the traditional process to one of "means-ways-ends." The emphasis is on improvements that "bubble up, in entrepreneurial fashion, from the lower ranks." This approach does not, of course, replace the need for strategic thinking or guidance from the top. Instead it supplements it. And there are, of course, times when companies must change their objectives, and then senior management must manage the discontinuity without undermining lower-level managers. Hayes summarizes as follows:

> In short, the company doesn't first develop plans and then seek capabilities; it builds capabilities and then encourages the development of plans for exploiting them. Rather than trying to develop optimal strategies that assume a static environment, it seeks opportunistic improvements in a dynamic environment.
>
> Such a "reverse" logic tends to be most effective in rapidly changing competitive environments. Fixed objectives are likely to lose their attractiveness over time as the company and its competitive environment evolve. A common vision, however, will keep people moving ahead. (1986)

In recent years, Michael Porter at Harvard has had considerable influence on management thinking about how to create competitive advantage including the role of technology. According to Porter, technology can alter industry structure in several ways: it can create barriers to entry by altering economies of scale and changing capital requirements (for example); change the bargaining relationship with buyers through product differentiation and switching costs; change the bargaining relationship with suppliers; create opportunities for substitution; change the basis of rivalry among competitors by altering cost structures and restructure industry boundaries. Technology is a powerful competitive force because it pervades the entire value chain of the firm, the basis for Porter's analysis. "Technology affects competitive advantage if it has a significant role in determining relative cost position or differentiation." (1985, p. 169) The benefit of improving cost or product differentiation depends on the generic strategy of the company: whether it competes on the basis of low cost, product differentiation, or market niche. Strategy, Porter argues, is a question of concentrating "on those technologies that have the greatest *sustainable* impact" and which meet the following four tests:

1. The technological change itself lowers cost or enhances differentiation and the firms technological lead is sustainable.
2. The technological change shifts costs or uniqueness (drivers) in favor of the firm.

3. Pioneering the technological change translates into first-mover advantages besides those inherent in the technology itself.
4. The technological change improves overall industry structure.

Essentially, both Hayes and Porter argue from a naturalistic point of view. The technology should continuously be nudged towards the competitive goals of the firm, driven forward by low-level (as opposed to low-ranking) expertise and exploiting opportunities as they present themselves. Again, this does not mean that the company is without goal or direction.

This perspective is an important departure from the conventional wisdom. Conventional wisdom says that information technologies should be applied to a specific problem or opportunity. A feasibility study, normally including a cost-benefit analysis, determines whether or not the application is acceptable. This approach is more or less consistent with the goal-oriented planning process. But the opportunistic view of competition conjured by Hayes says, in effect, that a company should position itself to take advantage of potential opportunities. This implies that investments in technology may be necessary not only to accomplish specific tasks but also to provide general competitive capacity. The exact uses of this capacity will not be known in advance. In these cases, cost justification of the technology may not even be possible.

The Portfolio Approach to Technology Investment

Ultimately, managers must decide the level of financial commitment they are willing to make to information technologies and how the investment dollars will be spent. This is the cental issue of the position-taking phase in the Integrated Technology Management Framework. Here, I am proposing that these decisions be made using what might be called a "portfolio approach" to technology investments. The portfolio approach is an attempt to achieve a balance between long-term competitive capacity and short-term profit performance. The trade-offs made in any particular company will, of course, vary. No single guidance is available for judging how management must achieve the balance. Instead, the purpose of the portfolio approach is to make explicit what the trade-offs are. Managing a technology "portfolio" is really no different than managing an investment portfolio: the major dimensions are risk and return.

Managing investment portfolios entails making three general decisions. The first is the overall goal of the portfolio, for example, growth or income, or both. This goal is predicated on the needs of the portfolio's owner. The second decision is how aggressive or conservative the investments will be which is, of course, a reflection of the degree of risk assumed. Finally, specific choices about investments must be made. The latter step is an ongoing process as less attractive opportunities are weeded out of the portfolio and replaced by more attractive ones. Each of these decisions will be examined in terms of how it relates to technology.

Establishing a goal for the technology portfolio is less a question of achieving specific objectives and more one of developing the technology infrastructure's capacity. The capacity to be developed depends on the opportunities available to the firm. The technology assessment methodology covered in the previous chapter addresses how these opportunities can be evaluated taking into consideration new and emerging technologies, likely changes to industry structure, and various responses which could be taken to market conditions. The capacity to be developed also depends on the firm's baseline or existing situation. Capacity has two components: (1) general competitive capacity and (2) short-term and medium-term exploitation capacity.

General Competitive Capacity

General competitive capacity should be developed in response to long-term trends in the industry. In financial services, for example, a number of major changes are occurring which are altering the industry. The growing use of telecommunications and the rising importance of the Tokyo Stock Exchange are creating a global investment market. Deregulation has blurred distinctions between commercial and investment banks, and financial and nonfinancial institutions. Companies such as Sears and American Express have acquired investment banking concerns to exploit their extensive credit and customer bases. Deregulation has also created the potential for truly national banking and with it heightened competition for consumer loyalty. Coming from a traditional background of slow change and conservative practices, financial service companies today compete by producing a rapidly changing array of new products and services. The primary tool for creating these is information technology. This means that companies who wish to make the transition from old-line banking, investment banking, or insurance have had to evolve the general competitive capacity of their technology infrastructure.

The Travelers Companies provides one example. We have already mentioned that Travelers was chiefly noted as a Hartford-based insurance concern. Like most insurance companies, its computer information systems had been built to process large volumes of transactions, not for product innovation and flexibility. Management's first instinct was to scrap the existing systems and begin anew. But further investigation gave them pause: The company had a $200 million investment in software alone and scrapping this asset base would be difficult to swallow. On closer examination, they also found that the entire system was logically subdivided and reflected the sequence of events in writing an insurance policy. Major subsystems had been built over the years: data entry, editing, rating rules (or rules of product), policy assembly, and policy production (see Figure 6–1). Thus, rather than tackling the system as a whole, the technology management team began to nudge the infrastructure towards the general competitive capacity needed for a fast-paced, consumer-oriented financial services market. They began to think of the existing systems as a "data processing factory," an administrative factory for the production of financial services products.

Using this perspective clearly delineated the point at which Travelers began looking at its data processing function as a strategic tool rather than as simple transactions support. This change in perspective resulted in a change in tactics by the technology managers; having in place those elements of the system that continued to be of value, they began working on improving the components that could allow for the flexibility they needed. Because of the consumer orientation, many of the modifications focused on the points closest to the customer: data entry and policy production. Eventually, according to Travelers, through electronic media including video text the company will be able to deliver products anywhere, including the home. The entry point has received even more attention. Building on the spirit of taking the technology to the agents with on-line terminal access, the company has begun to experiment with taking the technology directly to the customer. One medium being tested is the electronic "kiosk" which uses interactive video disk technology. Through use of a touch-sensitive screen a potential customer can ask

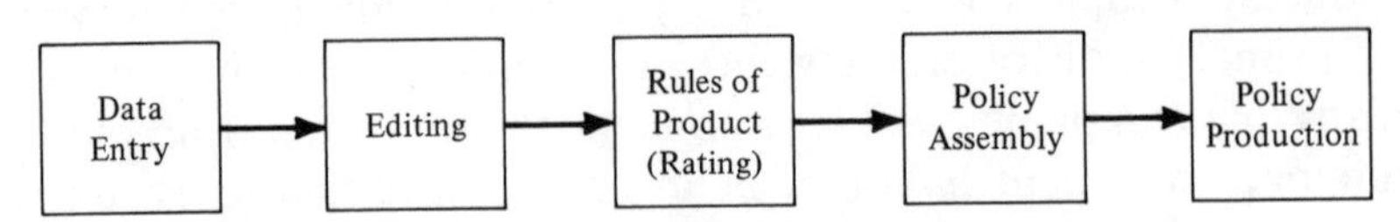

Figure 6–1. Data Processing Subsystems—The Travelers Companies.

questions concerning insurance or other financial services products. Actual rate quotes are available on the spot. In an early test in 1986, Travelers placed such kiosks in Denver supermarkets. Shoppers who interacted with the machine were asked to enter personal identification information which was later passed to a local agent. The agent then followed up on the electronic contact.

A second example comes from the publishing industry. McGraw-Hill has been one of the most aggressive publishers in moving towards the information industry concept. Joseph L. Dionne, Chief Executive Officer of McGraw-Hill, Inc., described how technology was changing the industry:

> Computer and communications technology alters the ways one can search out information, access it, edit it, and manipulate it, and tailor it to specific needs undreamed of not too many years ago. This new technology is changing the ways one can create new information products, reshape and enhance old ones, package and deliver them. For both information companies and their customers, it is even changing the concept of what information is.

In 1984, McGraw-Hill reorganized in recognition of the fact that major structural changes were reshaping the publishing industry. Specifically, the company sought to respond to expanding global markets for information, a growing marketplace for information in many forms and in many frequencies, and an increasing technological capability to meet these demands. Through strategic planning sessions, McGraw-Hill began to perceive early on that the growing complexity of decision-making throughout the business and professional communities was increasing the need for decision-support systems. In fact, according to their vision, many companies would eventually develop "corporate information centers." These centers would be facilities to garner information needed to support the company's internal functions. For McGraw-Hill and others the question became who was to be the vendor of corporate information centers and how would the information be delivered.

McGraw-Hill's strategy was organized around two major thrusts. First, the company was reorganized into market-oriented units that are "media blind." In the past, the company had been organized around products—traditional book publishing, on-line electronic data. "Media blind" means the company was less interested in the mode of information delivery and more interested in vertically integrated information markets. (Markets centered upon a single product or market niche). Thus, the second thrust was to evolve into a multimedia company which could leverage information resources. Initially, the company was concentrated on developing eight "platforms" to process and transform its print information into electronic information and disseminate it to users. Massive stores of information available for tailoring top specific user needs are maintained in large data bases which McGraw-Hill refers to as "data turbines." According to McGraw-Hill:

> Technology "platforms" is the term we use to describe the forms of technology we are using and developing in support of our products. These platforms consist of systems already in place and others we are developing with new software that McGraw-Hill has purchased. These platforms are available to all product developers (within the company) to help them launch new or enhanced products. Some platforms overlap. Some are most effective when used in combination.

The eight technology platforms include:

1. *Electronic composition.* Electronic composition permits automated text editing of diverse information. At Datapro, a subsidiary, McGraw-Hill processes all information about computers, communications, and office systems through an editing system called "STEX." These and similar composition systems have two major advantages. They improve editorial productivity because employees need to enter the information into a computer only once for editorial preparation and for later page printing. These also convert the information into a form suitable for later electronic distribution.
2. *Timesharing.* McGraw-Hill uses timesharing, the second platform, to provide on-line access through communications networks to make large data bases selectively available to remote users with personal computers. Customers in the construction industry—one of the vertical information markets—can receive a five-year construction market forecast by quarter as well as associated economic and demographic data.
3. *Software publishing.* Software publishing involves packaging data with programs and transferring them on disks for business people to use on miocrocomputers in their offices. For example, McGraw-Hill developed a software program that utilizes construction cost data. The data comes from McGraw-Hill's own large data bases for on-line customers. These packages

therefore encourage customers to use McGraw-Hill's specialized data bases even more. This third platform also makes use of the most advanced technology available for electronic products on disks. McGraw-Hill has developed a product selection program based on the physical and technical characteristics of building products. Their system uses a compact disk with read-only memory (known as CD-ROM) to fit the characteristics of all products in nearly 6,000 catalogs into one disk. Through this program, construction designers and contractors are able to find the products they need and the identities of manufacturers represented in the files.

4. *Real-time feed.* The fourth "platform" is real-time data feed which makes new information available as soon as it is created. In construction, McGraw-Hill makes available immediate news of major construction projects worldwide as well as financial and currency markets. Service will eventually include construction cost data, economic trend data, and an interactive trade service allowing users to find project partners, financing, and services.
5. *Computer conferencing.* Special software enables a main computer to handle numbers of simultaneous conferences. Each conference has timesharing participants interacting at the same time. Computer conferencing permits users to organize exchanges on topics of current interest. The software provides statistics showing how popular each conference is so that decisions can be made about which ones continue and how to organize new conference topics.
6. *Full-text retrieval.* The sixth platform is full-text retrieval which permits a customer to search through an entire data base for all information related to his interest. Full-text retrieval requires conversion of all printed material to electronic data, ample computer space to store the entire data base so it is accessible for on-line queries, and considerable computer power to conduct searches of the complete data base.
7. *Videotex.* The seventh platform is videotex, a system of hardware and software that combines textual information with graphics and full-page color in formats to provide interpretation through creative display. Applications which involve location maps, site plans, advertisements, and similar graphic-oriented material are possibilities.
8. *Broadcasting.* Through a partnership, McGraw-Hill had begun

to broadcast general business, stock investment, and economic news to owners of desktop computers by way of satellite and cable television lines. The service provides a one-way data stream and a software program that enables recipients to select the categories of news they want to view.

There are several striking similarities between what McGraw-Hill and The Travelers Companies are doing with technology. Like Travelers, McGraw-Hill has been confronted with major changes in its traditional business. Just as information technologies facilitated the integration of insurance with other financial services products previously considered to be outside the industry's boundaries, so was traditional publishing integrated with other forms of information delivery to create a new industry, the "information industry." Other major players in this industry include Reuters, Dun & Bradstreet, Quotron, TRW, and Dow Jones. Travelers and McGraw-Hill also found themselves catering to more and more specialized consumer categories. In this sense, technology facilitates finding and servicing specific market niches. And both companies also found that they were producing more products and at a greater rate.

Two important conclusions derive from these similarities. First, technology in both organizations played some role in all three generic strategies: lowering costs, differentiating products and services, and creating market niches. Second, and more important, both companies focused their efforts on creating a competitive technology infrastructure. What Travelers calls a "data processing factory," McGraw-Hill calls data turbines and platforms. Although there are slight differences, the concept is similar: to provide general competitive capacity appropriate to the direction of the industry.

The investment decision related to general competitive capacity is basically long-term; the purpose is not simply immediate return. The criterion for success is to be able to answer the following question in the affirmative: Will the company have the appropriate technological capacity to compete in the industry five to ten years into the future or more?

We have already seen that developing technology systems in the factory and office sometimes require years. Thus, the question of how to create competitive advantage with technology is perhaps best answered by saying that the victor will be the company that knows where the industry will be a few years out and begins now to make sure the general technology capabilities will be in place.

Short-Term and Medium-Term Exploitation Capacity

The second major capacity decision relates to short-term exploitation of the technology infrastructure's capabilities. Assuming that the appropriate infrastructrue is in place, then individual business units within the organization can modify, recode, or otherwise manipulate the elements to solve relatively unique and often sophisticated problems in a fairly short period of time. Both Travelers and McGraw-Hill demonstrate this capability. Another example is afforded by Dun & Bradstreet, one of the players in the electronic information industry which produces credit and miscellaneous business information. Like McGraw-Hill, Dun & Bradstreet maintains large data bases from which it tailors information to specific market segments. The data bases were built on the credit history of seven million businesses plus marketing and other data for 75 million U.S. households. The company invests substantially in software to manipulate the data; in other words it has developed short-term capacity to exploit marketplace opportunities which can be serviced by its technology infrastructure. In the words of one Dun & Bradstreet executive:

> We've been able to take an existing database, reformat it, spin it around, and develop a whole slew of new products for our customers. Ten years ago, Dun & Bradstreet credit reports were as uniform as Model T Fords. Today they challenge Heinz for variety. (Gannes, 1985, p. 39)

Applications of short-term capacity may take one or two general forms. First, applications may be developed by end users. End users are essentially nontechnical employees—financial analysts, accountants, managers, marketers, salespeople—who use information technology to facilitate their tasks. End users generally regard the technology infrastructure as an information utility or resource to be tapped as needed. Any incremental investments needed in equipment, training, or related items are likely to be relatively small and should be budgeted as an expenditure of the end user's business unit. In most cases, end-user applications development would take less than one year. Indeed, the needs at this level are typically so specific that any attempts to link them directly to the larger organizational mission would be academic at best. It is simply a matter best left under the control of the local manager of the business unit. However, this approach should not downplay the importance of the technology infrastructure in servicing end-user needs and management should take special care to obtain continuous feedback to ensure that these needs are being met. If not, alteration of the infrastructure may be indicated.

The second form of applications is the familiar systems development project which would be appropriate for large, complex, and relatively more permanent exploitations of the infrastructure. The basic difference is that management would typically budget for the project, assign a development team, and implement over a one- to three-year time frame. Thus, this type of application would require a joint effort between the business unit and the technical information systems people. Because each project is now competing for limited corporate resources, some means of evaluating and ranking opportunities must be employed. The appropriate criterion is contribution to *critical success factors.*

The notion of critical success factors (CSFs) was developed several years ago and refers to those few key areas of activity in which favorable results are absolutely necessary for the business to reach its goals. According to John F. Rockart:

> . . . critical success factors are areas of activity that should receive constant and careful attention from management. The current status of performance in each area should be continually measured, and that information should be made available. . . . (1979, p. 85).

Critical success factors are not goals but important avenues in reaching goals. An automobile manufacturer might have earnings per share, ROI, market share, and new product success as goals, but the factors critical to success may be styling, quality dealer system, cost control, and meeting energy standards. Information systems applications should focus on these factors. According to D. Ronald Daniel:

> . . . a company's information system must be discriminating and selective. It should focus on "success factors." In most industries there are usually three to six factors that determine success; these key jobs must be done exceedingly well for a company to be successful. (1961, p. 116)

An important realization is that critical success factors change as the company's situation changes. Thus, these factors must be continually redefined and clearly articulated and technology projects restructured to reflect the evolution. Many managers find this frustrating but it is necessary and the natural order of things.

Figure 6–2 summarizes the relationship between the technology infrastructure, general competitive capacity, and short- and medium-term exploitation of that capacity. Through the technology assessment process the company gathers both internal and external intelligence on the company's current capabilities (baseline), how technology will develop and affect industry structure, and various positions the company might take in response. From this and information collected on the general business environment the company reaches some measure of long-term business goals. Investments in technology must then address two levels of issues. First, the infrastructure must evolve to provide general competitive capacity five to ten years hence. This long-term capacity development positions the company vis-à-vis other competitors in the industry and positions it to reap short-term benefits. Second, assuming that the appropriate infrastructure is in place, the company exploits short-term opportu-

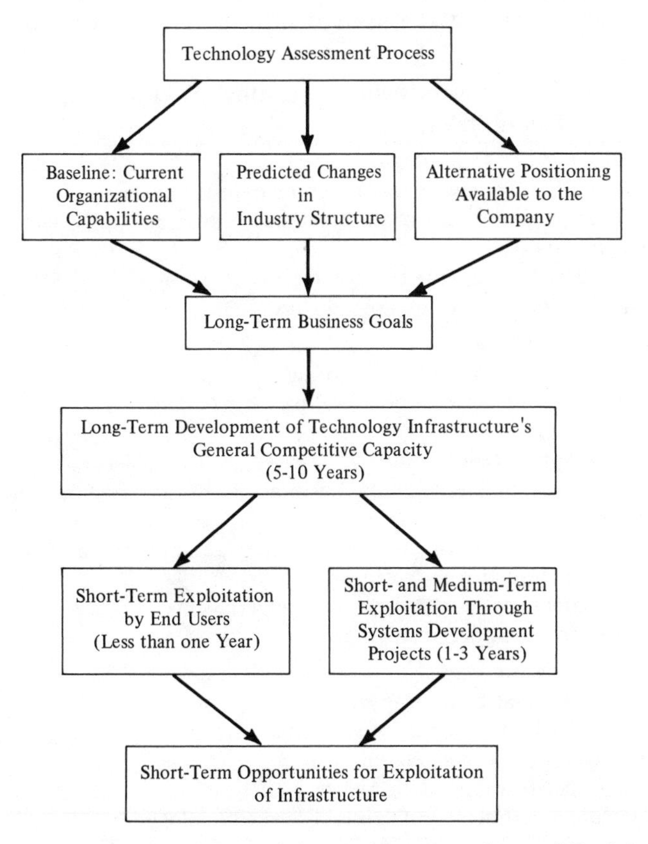

Figure 6–2. Infrastructure, General Competitive Capacity, and Exploitation.

nities through applications developed by end users or short- to medium-term opportunities through system projects of one to three years or more in duration.

Offensive or Defensive Postures

The second major portfolio decision is whether the company should take a leadership position by aggressively using technology to implement business strategy, or whether it should remain in a defensive posture, traveling more or less as a pack with other competitors. Again, Michael Porter adequately treats this question elsewhere, arguing that three factors determine whether a company should seek technological leadership: (1) the sustainability of the technological lead, (2) first-mover advantages, and (3) first-mover disadvantages. These factors are summarized in Table 6–1.

Anecdotal evidence suggests that companies that do well at managing technology thrust themselves into an aggressive posture. The

Table 6–1. Factors Influencing Technological Leadership or Followership.

Sustainability of the Technological Lead
—Was the technology developed inside the industry or outside it?
—Is there a sustainable cost or differentiation advantage in the technology development activity?
—Does the firm have unique technological skills vis-à-vis competitors?
—How quickly does the technology diffuse?

First-Mover Advantages
—Establishing a reputation as the pioneer or leader
—Preempting an attractive product or market position
—Creating switching costs
—Gaining unique access to brokers, distributors, or retailers
—Gaining a cost or differentiation advantage from experiencing a proprietary learning curve
—Enjoying favorable access to facilities, inputs, and other scarce resources
—Defining standards for technology
—Securing patents or special status with governments
—Enjoying early temporarily high profits

First-Mover Disadvantages
—Bearing substantial pioneering costs
—Bearing the risk of uncertainty over future demand
—Being vulnerable to changing buyer needs
—Changing specifications of current technology
—Obsolescing investments in the established technology

Source: Porter, 1985, pp. 182–191.

Travelers Companies, as already noted, maintains a large staff of computer scientists who work at applying information technologies to financial services. Their efforts gain Travelers many of the first-mover advantages listed in Table 6–1. Travelers cultivates an image as a leader in its field, the company has input into vendor development of computer products, and substantial efforts are made to better position its marketing of products and services. For Travelers this creates some advantage, although it is difficult to sustain. The types of technologies Travelers employs in its business are broadly available and therefore accessible by competitors. The remedy is to continue aggressively developing the technology.

Once competition with technology becomes keen, most firms have little choice but to follow the lead. Countless managers will, when pressed on the issue, concede that many investments in technology are made simply because competitors have made the investment. This is to be expected at least to the extent that several competitors make assessments of the long-term competitive environment (including technology's impact) and reach similar conclusions. Thus, each company's technology infrastructure will be targeted to the same general competitive capabilities.

The important issue in deciding on an offensive or defensive posture may not be the jockeying for position between individual competitors but how an offensive posture by any one firm or group of firms will restructure the industry as a whole. Again, automated teller machines (ATMs) provide an excellent example. ATMs were originally introduced by the largest New York banks partially as a means of achieving and sustaining first-mover advantages. Specifically, Citibank's concept was to create a proprietary network that would create switching costs for consumers and further establish the bank as a leader in the industry. Switching costs would be created, so the logic went, because consumers would become familiar with the system and would not want to learn another. The expense of developing an ATM network is so high that many thought smaller banks could not enter this arena. But smaller banks banded together and created shared ATM networks such as NYCE and CIRRUS. By sharing development costs and using a different tactic (i.e., shared versus proprietary network) the smaller banks were able to get past the barriers.

Over the past few years, Citibank's market share of depositors has increased several points and this gain is attributed in part to ATMs. However, it is ultimately very difficult to measure the impact of

ATMs in helping gain market share. And for the impact that did occur, was the investment well spent? Many argue that service to depositors has improved; queues are shorter, service hours are longer. While this may be true it leaves unsatisfied the question of what it did to improve the industry. If all banks are now spending more money to maintain the same market share or only a marginally greater share, where is the benefit to any single competitor within the industry? In Chapter 3, I noted that ATMs do not cost justify on productivity grounds alone (i.e., banks are beginning to charge for the service). And, although ATMs may be an important conduit for selling a broad range of financial service products, no one has a clear understanding of when and how or even if that process will produce acceptable returns.

Little in the way of a definitive or acceptable answer is available on this issue. However, the uncertainties raised by being the first-mover in technology-based competition provide a powerful incentive to the would-be offensive competitor to carefully assess what forces will be set in motion for the industry as the result of strategy. What goes wrong with some offensive strategies is that companies mistakenly asume that they can maintain exclusive control of a specific technology. Experience proves that this is very difficult to do. Sustaining an offensive lead can be extremely costly.

Profiling the Technology Investment Portfolio

In the final analysis, managers must periodically determine what the commitment of expenditures will be to a specific technology, and the sum or whole of these decisions constitutes the technology investment portfolio. Senior management must be able to assess what the overall contribution of the portfolio means to the firm. This, of course, is the basis for having reasonable expectations of the technology. For purposes of better understanding the portfolio, a technology investment profile should be kept updated as often as necessary. The exact method by which the profile is developed will vary from company to company; much of the contents is subjective in nature. But, a general model of the technology investment profile is presented in Table 6–2.

The profile is meant to measure two major characteristics of any portfolio: risk and return. Risk, with respect to information technologies, was defined by F. Warren McFarlan of Harvard University as follows:

Table 6–2. Technology Investment Profile.

DIMENSIONS	RISK	RETURN
1. Past impact of information technologies on strategy	Lower	Higher
2. Anticipated impact of information technologies on future strategy	Higher	Lower
3. Nature of the change (P2 versus P1)	Higher	Lower
4. Orientation towards products and services versus internal operations	Higher	Lower
5. Offensive versus defensive posture	Higher	Lower
6. Active technology assessment	Lower	Higher
7. Upstream technology development versus downstream development	Lower	Higher
8. Size of investment	Higher	Lower

> By risk I am suggesting exposure to such consequences as: Failure to obtain all, or even any, of the anticipated benefits. Costs of implementation that vastly exceed planning levels. Time for implementation that is much greater than expected. Technical performance of resulting systems that turns out to be significantly below estimate. Incompatibility of the system with the selected hardware and software. (1981, p. 143)

Return does not refer to magnitude as much as to predictability and measurability. A predictable and measurable return is not necessarily better than an unpredictable and unmeasurable one. To the contrary, I have already suggested that the process of building the general competitive capacity of the infrastructure escapes familiar cost justification techniques. Rather, the issue is when one is in uncharted territory and when one is on firm ground. A particular company must consider the mix of returns expected from its portfolio. The desired mix will depend on what immediate returns must be generated to satisfy financial constraints faced by the firm. By artfully constructing a portfolio which includes at least some technology projects with a high current yield in terms of measurable benefits, a balance can be struck between short-term and long-term benefits, between short-term exploitation of the current technology infrastructure and building general competitive capacity.

Eight dimensions that form the profile of the portfolio (see Table 6–2) are discussed here.

Past Impact of Information Technologies on Strategy

We have already discussed how companies that have reaped substantial rewards from past automation are more likely to create a positive, reinforcing technological climate within the firm. The likely result is that management has a greater understanding of the benefits that technology produces and overall a more realistic level of expectations. Thus, all other factors being equal, the risk of a technology investment not yielding anticipated benefits, performing poorly, or being improperly implemented is lower. The return in terms of predictable and measurable benefits is higher.

Throughout the book, I have used both airlines and financial services as examples of industries in which technology has been important strategically for decades. Industries with less experience, especially those wading into technology for the first time, are likely to incur much greater risk and be less successful predicting the outcomes of their investment decisions.

Anticipated Impact of Information Technologies on Future Strategy

Based on the technology assessment procedure already described, management may have determined that the investment has important implications in terms of the firm's ability to compete in the future. The reverse of this situation is one where the investment might be profitable, but lacking in any particular strategic value. For example, a financial analysis unit is automated primarily to obtain P1 type productivity gains, short-term and immediately realizable. Typically, as the importance of an investment to future competitive ability rises, so does risk. The consequences of not obtaining desired benefits or implementing the system on time would be greater because of its strategic importance. By the same token, an investment for future strategy purposes would tend to be less predictable and measurable in terms of return. As one looks farther into the future to general competitive capacity, the return is certainly less predictable than when making investments in technology for the purposes of short- or medium-term exploitation.

When the first and second dimensions are combined, a "strategic grid" emerges similar to that developed by McFarlan, McKenney, and Pyburn (1983) shown in Figure 6–3. The first dimension, past impact on strategy, is depicted on the vertical axis and, the second, future impact, on the horizontal. Four characteristic types of firms

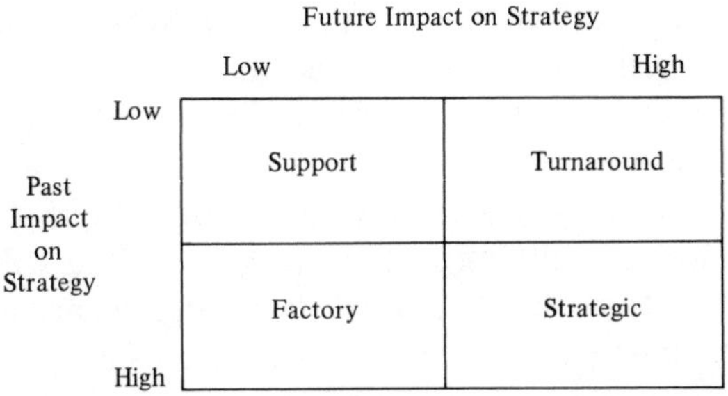

Past Impact on Strategy	Future Impact on Strategy: Low	Future Impact on Strategy: High
Low	Support	Turnaround
High	Factory	Strategic

Figure 6–3. Strategic Grid.

Reprinted by permission of the Harvard Business Review. *An adaptation of an exhibit from "The Information Archipelago—Plotting a Course" by F. Warren McFarlan, James L. McKenney, and Philip Pyburn (January/February 1983).*

emerge: support, factory, turnaround, and strategic. A "support" firm neither depends on past (existing) information systems as a critical component of its success nor does it anticipate doing so in the future, although it may still have a substantial investment in technology. According to McFarlan and his colleagues, certain large manufacturing companies may fit into this category. Technology may be used for accounting, payroll, and inventory, among other applications, but these have little strategic significance. The risk profile for support firms is somewhat mixed. Because past applications have not been strategically important, experience has been less. Future investments therefore carry a higher risk. But this is offset by the fact that future applications are not strategically important. The return profile is the opposite. Since past experience has not been critical, senior management has had little opportunity to hone their understanding of the benefits and the lack of future impact lowers their motivation.

The second type is "factory." While the factory type has no particular impact on information technology for future strategy, it differs from the support firm in that it is highly dependent on existing technology. Here risk and return are generally consistent. Risk is lower because the learning curve has been essentially scaled. Returns are predictable and measurable because future investments are based on past experience.

"Turnaround" firms, the third type, represent substantial uncertainties. Here the strategic impact of past technology has been low, so that risk is higher and returns less well understood. Likewise the future strategic impact of technology reinforces higher risk and lower predictability of returns. Companies that fit this mold are normally at greatest risk. Essentially, their managers must make important strategic changes with technology with no relevant background experience.

The fourth and final type is "strategic" firms. For them, technology has been strategically important in the past and will continue to be in the future. Like the "support" type, signals are mixed on risk and return but past experience weighs in their favor.

Taken as a whole, the strategic grid shown in Figure 6–3 provides a simple ranking of investment types that provides senior management with some insight into the appropriate level for its own involvement and commitment to resources for technology assessment. In order of importance, they are: (1) strategic, (2) turnaround, (3) factory, and (4) support. The higher the order, the greater the importance (or risk) to the company. It follows that senior management has a greater role to play, that having a technology person on the executive staff makes more sense, and that the firm should be much more aggressive in assessing new and emerging technologies.

Nature of the Change (P2 versus P1)

The third dimension affecting risk and return is the nature of the change implied by the investment. If the change is significant structurally, implying P2-type productivity gains, then risk is likely to be higher and return less predictable. The reasons are that P2 productivity is relatively untested and it requires tampering with major organizational change as well as technical considerations. P2 changes the way in which business is done. P1 productivity change requires no such restructuring of the business and is, therefore, less risky and more predictable where returns are concerned.

Orientation Toward Products and Services versus Internal Operations

Information technologies can be oriented toward the product or service in the sense that a new or changed product or service emerges; in other words, the technology has a visible impact on the competitive marketplace. Because customer or market reaction is involved, investment decisions in product or service-oriented technologies are inherently more risky and carry less predictability on returns. The reverse situation exists when the investment decision relates to internally oriented technologies; risk is less and return is more predictable and measurable.

Just as the first and second dimensions can be combined to produce the "strategic grid," the third and fourth dimensions can be combined to form what Benjamin, Rockart, Morton, and Wyman (1984) referred to as the strategic opportunities framework. This framework, liberally adapted, is shown in Figure 6–4. The combination of the two dimensions yields four competitive positions. Each position has a different degree of associated risk and return.

P1 CHANGES IN INTERNAL OPERATIONS

One position involves no basic change traditional business practices, a P1-type productivity strategy, and internal operations. An information system supporting personnel or the use of desktop computers to assist with financial analysis are examples. Both the P1 approach and internal orientation involve a relatively low degree of risk and highly predictable return.

P2 CHANGES IN INTERNAL OPERATIONS

The second position also confines itself to internal operations but involves P2-type structural changes. Westinghouse Corporation, for example, has since 1980 mounted an extensive effort to reshape white-collar work through various technology applications such as electronic mail, teleconferencing, voice messaging, robot mail delivery, personal computing, and more. Although the impact of these changes may be felt by parties external to the company, the target is improving internal productivity through significant structural

	Internal Operations	Competitive Marketplace
Traditional Business Pratices	P1-Oriented Productivity Strategy	P1-Oriented Productivity Strategy
Significant Structural Change	P2-Oriented Productivity Strategy	Industry Structure or Information Industry

Figure 6–4. Strategic Opportunities Framework (adapted).

Reprinted from "Information Technology: A Strategic Opportunity" by Robert I. Benjamin, John F. Rockart, Michael S. Scott Morton, and John Wyman, Sloan Management Review *(Spring 1984), pp. 3–10, by permission of the publisher.*

change. Both risk and return are mixed for this position. The P2 approach introduces a somewhat higher risk and less predictable return.

P1 CHANGES IN THE COMPETITIVE MARKETPLACE

The third position is product and service oriented although no major structural change is involved. A P1 approach is used. Credit card operations are an example of this position. Information technologies are continuously applied to improving credit card services to customers by offering special services, sales promotion, theft protection, and similar benefits. Risk and return tend to be mixed. Although a P1 approach lowers risk and makes return more predictable, the impact on the customer raises risk and lowers predictability. A minor computer error in calculating sales tax on a special merchandise offering could have large-scale adverse consequences simply because of the number of customers involved.

P2 CHANGES IN THE COMPETITIVE MARKETPLACE

The fourth position is the most risk prone and least predictable since it combines both changes to the business structure and an orientation to the marketplace. McGraw-Hill, for example, is using technology to restructure its business (and the industry) in a way that is highly visible to the customer.

Offensive versus Defensive Posture

The fifth dimension of the technology investment profile is whether a company takes a leadership position with technology or remains a follower. The leader is the first to mount the learning curve in the application of a technology and the surprises it holds create a substantially higher risk and lower ability to predict returns.

Active Technology Assessment

The sixth dimension is how active the company is in assessing new and emerging technologies. Aside from gathering intelligence, the technology assessment function can play an important role in raising the overall knowledge of technology within the organization. Even without experience, the informed decision maker is better off than the uninformed one. The assessment function itself is important because it institutionalizes the collection and flow of information about technology. It helps remedy the impressionistic, unorganized, and incomplete knowledge of technology that currently characterizes management thinking. For these reasons, active technology assessment reduces risk and increases the predictability of return. Indeed, of all the options available to American managers today to improve technology management, one of the most readily implementable actions is to organize and strengthen the company's technology assessment function.

Upstream Technology Development versus Downstream

Closely associated with an active assessment function is the active involvement of the company in upstream developments with technology through research sponsorship, joint ventures, and other vehicles. The more active the company becomes as a participant, the greater its overall understanding of the fundamental nature of the technologies and their implications for business. Therefore, upstream participation is likely to reduce the risk and improve return predictability. Generally speaking, most American companies view themselves as passive recipients (downstream) in technology. What business does a bank, investment house, or airline, for example, have in upstream development activities? Yet the growing central importance of information technologies to business strategy has so increased the stakes that more and more companies are moving upstream.

Size of Investment

The eighth dimension is the size of the investment being contemplated. Size and complexity are synonymous. As a project becomes larger, more staff and time are needed for implementation, and the complexities grow exponentially. This is characteristic of nontechnology as well as technology investments. Greater size and complexity obviously increase risk and lower predictability of returns. A project that seriously overruns budget or fails to generate benefits because of time delays can become a burden rather than an advantage to the business.

The foregoing discussion of risk and return behavior for each of the eight technology investment dimensions has been presented in terms of general tendency. In certain situations, the effect could be reversed. The impact of technology on strategy may have been high in the past, generally lowering risk and improving predictability of returns, but the past experience could be irrelevant to the current investment consideration or bias the decision in a dysfunctional way. This would increase risk and lower return predictability. Therefore, profiling risk and return for individual investments along the dimensions proposed must be done with care and judgement. According to F. Warren McFarlan:

> In addition to determining the relative risk for single projects, a company should develop an aggregate risk profile of the portfolio of systems and programming projects. While there is no such thing as a correct risk profile in the abstract, there are appropriate risk profiles for different types of companies and strategies. For example, in an industry that is data processing intensive, or where computers are an important part of product structure (such as banking and insurance), managers should be concerned when there are no high-risk projects. In such a case, the company may be leaving a product or service gap for competition to step into. On the other hand, a portfolio loaded with high-risk projects suggests that the company may be vulnerable to operational disruptions when projects are not completed as planned.
>
> Conversely, in less computer-dependent companies, [information systems] play a profitable, useful, but distinctly supporting role, and management often considers the role appropriate. In such cases, heavy investment in high-risk projects appropriately may be smaller than in the first type of company.
>
> Even here, however, a company should have some technologically exciting ventures to ensure familiarity with leading-edge technology and to maintain staff morale and interest. Thus the aggregate risk profiles of the portfolios of two companies could legitimately differ. (1981, p. 144)

Concluding Thoughts on Position-Taking

Two concluding thoughts are presented as something of a postscript on the position-taking phase. First, no one approach to technology utilization exists. There is not, as some think, a correct and incorrect way to employ computer-based technologies. Indeed the whole beauty of more recent developments is that technology can be molded to the organization's form and function. Form and function should, of course, be based on the nature of the organization's mission. Thus, the litmus test of success with technology is whether the company is accomplishing its mission. It is in this context that one can speak of "appropriate" technology rather than correct technology.

The second thought revolves around the question: "What should technology be appropriate to?" The obvious answer is that it should serve the company's business goals. But this assumes management has well-developed and articulated goals. Companies with a strong sense of purpose and direction typically find technology much less daunting than would otherwise be the case. Companies with weak objectives often find that technology is only a more powerful vehicle for going nowhere in particular. As an example, a major international freight-forwarding company began to automate its administrative process some years ago. Growth in the closely held firm had been rapid, and due to the entrepreneurial background of the owners, rigorous analysis of segment profitability was totally absent. The owner/managers could not tell whether any particular division was making or losing money. Indeed, it all seemed academic anyhow since lucrative contracts with Middle Eastern oil-producing nations guaranteed the company would make money in spite of itself.

The most visible and troublesome problem faced by the company's management was an inadequate inventory system. Customer shipments were forever getting lost and clerks would have to make physical searches of the warehouse until the lost freight was found. The company invested heavily in new computer equipment and began building an expensive inventory tracking system. This proceeded smoothly until some of the lucrative contracts were lost. The added expense of computing (not yet producing benefits) quickly put the company in the red for the first time. Consultants who were called in to examine the situation pointed out that while the new inventory system might indeed solve a thorny problem for the company, it did not address critical problems. In other words, the inventory system could not make a material contribution to the company's profitability. Work on the inventory system was halted while the company hastily installed off-the-shelf software to improve segment profit analysis. The results of these analyses helped management focus their energies on areas where profitability could be improved.

When technology investment decisions have been made, a company has charted its course and the position-taking phase comes to a conclusion. Many managers also believe that their responsibilities are largely concluded at this point. But, the third phase—policy formulation—remains and ranks in importance as at least equal to the first two. Phase III addresses the need for senior management to develop and promulgate policies necessary to steer the organization and workforce, and perhaps even external parties, in directions consistent with its technology-based strategy.

PHASE III

Policy Formulation

7

Technology and the Organization

The Purpose of Organization

One of the basic premises of this book and the foundation on which the Integrated Technology Management Framework rests is that answers to questions about managing technology are best sought by examining traditional concepts. This does not mean that management's thinking should not be challenged or that new ideas should not be introduced, nor does it mean that the management process itself will not evolve. Certainly it must. But when viewed in a vacuum, technology presents us with fantastic visions of what the future *could* be and with the application of imagination we find ourselves contemplating what it will be. Nowhere has this predilection for futurism been so vigorously exercised as when speculating on technology's impact on the organization.

One example is the popular conception that information technologies will eliminate bureaucracy within organizations. According to this view, advances in information technologies coupled with a growing need to be more responsive to rapidly changing markets will cause the hierarchy to wither away, leaving small entrepreneurial teams that will network to accomplish tasks as the need arises:

> We see a revolution on the horizon that holds far-reaching implications for the American corporation. A combination of forces—from the rapidly changing business environment to the new work force to astonishing advances in technology—is forging a breakdown of the large traditional, hierarchial organizations that have dominated in the past. We think that this dismantling will result in highly decentralized organizations in which the work of the corporation will be done in small, autonomous units linked to the megacorporation by new telecommunications and computer technologies. This change can turn us all into entrepreneurs and in the process will transform the role of middle management. Motivation will come from the opportunity to accomplish complex tasks in an animate, relatively simple work environment. We won't waste time and energy worrying about how to climb the corporate ladder because there won't be one. Most middle management rungs will be replaced by mechanisms of social influence by emphasis on culture. We see it as a no-boss business. We call it the *atomized* organization. (Deal and Kennedy: 1982, p. 177)

Such descriptions ignore certain basic realities concerning the importance of hierarchy and even bureaucracy in organizational life. Hierarchy, for example, serves several useful purposes including:

1. *Task Specialization.* Hierarchy provides the means for specialization and development of expertise within organizations as well as the means for coordinating different roles and preventing them from interfering with each other. Hierarchy provides direction in terms of which individuals will perform what tasks.
2. *Communications Channels.* Hierarchy serves as a means of *structuring* the channels of communications. A specific question regarding costs of operating a telecommunications link between the United States and Japan, for example, might first be referred to the head of the telecommunications unit, who in turn would refer it to a specialist within the unit.
3. *Integration.* Hierarchy also serves an important purpose in binding the organization together and instilling discipline through objective setting, individual rewards, and similar mechanisms. Indeed, as an organization becomes more specialized, the problem of coordinating the whole becomes critical. Specialization cannot occur without hierarchy.
4. *Rights of Individuals.* Hierarchy provides a means to define rights and responsibilities of individual members and in this sense puts important checks and balances on the use of power.

> Readers can easily imagine that an organization of "small, autonomous units" will quickly begin to reorganize itself along the lines of coalitions and similar measures which build power bases. To believe otherwise is to blatantly ignore certain commonly understood human motives.

Most people understand that bureaucracies are also designed to accommodate the "average" person, not the human of above-average intelligence, initiative, and wisdom. But most employees are average people. Bureaucracies also act to bind members together, eliminating unwanted external influences. While bureaucracies often fall short of expectations, it is because of the variability of the people who populate them. Bureaucracy and hierarchy are endemic to American business enterprise and to propose removing them is to propose fundamental change to a principle underlying the American economy. According to Charles Perrow, sociologist and long-time student of organizations;

> Alternative forms of organization are being tried. Nevertheless, they have yet to do more than to humanize rigid bureaucracies and make them more adaptive to changes. They have not seriously challenged the wage system, which takes in about 85 per cent of the gainfully employed who must work for someone else. Nor have they fully rejected the notion that the resources of the organization belong to the owners or private firms or the officials of state and voluntary organizations, rather than to all members. Without fundamental changes in the wage system and ideas of ownership, alternative forms of bureaucracy are likely to be expensive, unstable, and rare. (1986, p. 4)

"Bureaucracy" and "hierarchy" are more often than not used in the pejorative sense in current conversations. Ours is the age of the entrepreneur and innovation. Yet, this perspective reveals a certain ignorance of what these terms represent operationally. It also ignores the fact that bureaucracy and hierarchy are important responses to many business problems. It is not to say that bureaucracy is without its shortcomings. Charles Perrow, again:

> Critics usually attack bureaucracy for two reasons—it is unadaptive, and it stifles the humanity of employees. Both are legitimate criticisms to a degree. But what the first avoids noticing is that another description of unadaptiveness might be stability, steadfastness, and predictability. If we want a particular change and fail to get it, we blame the unadaptive bureaucracy. If it changes in ways

> we do not like, though, we call for stability. The second criticism—that bureaucracies stifle spontaneity, freedom, and self-realization—is certainly true for many employees, but unfortunately, since they do not own what they produce and must work for someone else, expressions of spontaneity and self-realization are not likely to result in better goods and services for consumers. We have constructed a society where the satisfaction of our wants as consumers largely depends on restricting the employees who do the producing." (1986, p. 5)

Despite the central importance of bureaucracy and hierarchy, organizations can be varied in important ways. Management can be centralized or decentralized, emphasis can be on control or innovation, cultural ethos can be strong or weak, responsiveness can be quick or slow. These variations are specific responses to the problems and opportunities faced by the organization. They exist for a purpose. The organization tends to evolve to mirror its environment. In that sense, one management dictate has always been to shape and prepare the organization to meet the challenges before it.

To say that information technologies will create new organizational forms is to commit the same "sins" with technology as in the past when totally integrated management information systems were foisted upon management. Technology should not drive organizational form or anything else. The organization should be designed to fit the job at hand and information technology systems must be designed consistently with the organization's form. In this we are fortunate, for as we saw in Chapter 3, the growth in the variability of technology away from central data processing to the kaleidoscopic distributed processing environment has created an almost limitless range of options for building a technology infrastructure appropriate to a particular organizational form. This is senior management's objective: to align or match organizational form to the tasks it must accomplish and then to align the technology infrastructure. Organization follows mission; technology follows organization.

The primary tool that management has for acomplishing the alignment task is policy, which is Phase III of the Integrated Technology Management Framework. In the remainder of this chapter, the task of formulating policy will be considered in three areas related to the organization as a whole. These are: (1) organizational culture, (2) organizational structure, and (3) management style.

Technology and Organizational Culture

Culture is a sweeping term which encompasses many aspects of the organization. It embodies corporate philosophy and "superordinate goals," the overarching values that guide the business's activities. No contemporary discussion of managing technology would be complete without considering the substantial movement to reorient American corporate cultures. The drive behind the movement is the emergence in the international marketplace of countries such as Japan which hold a different set of values about competition, many already discussed in foregoing chapters.

One of the figures in this transformation of American corporate culture is W. Edwards Deming, originally a statistical quality control consultant who later became known for his management philosophy. Deming played an important role following World War II in helping the Japanese establish new principles of management based on quality and productivity. His influence is considerable in certain Japanese circles and indeed he has received numerous awards. According to Deming and others, long-term business success depends on building markets by *satisfying customers* and through *continuously improving the products and services.* By focusing on quality, Deming argues, the rest will pretty much take care of itself, as depicted in Figure 7-1. A causal chain of events is set up so that improving quality in the process as well as products and services causes costs to decrease which in turn leads to improvements in productivity. Lower costs and better product and service quality help capture market share. The result is success in business and more jobs.

This varies considerably from the traditional American approach which views quality as a trade-off from a cost-benefit perspective. If the cost to improve quality will not increase revenue so that profit is greater, then quality is not pursued. In production, the same is true. Production processes are designed for an "acceptable" level of defect. Pushing production further towards zero defects, according to the American view, causes large increases in marginal costs.

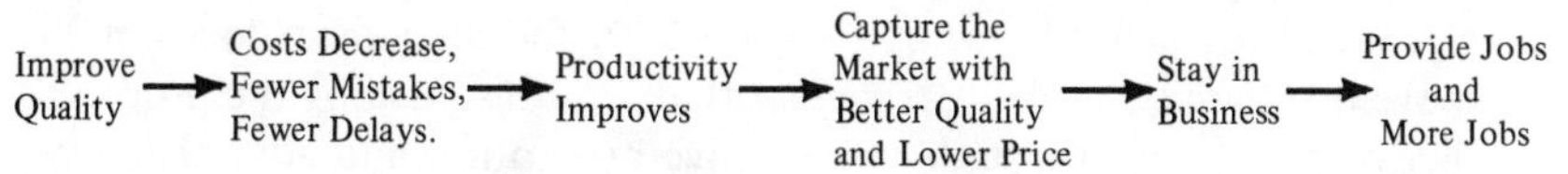

Figure 7-1. Deming's "Chain Reaction" (adapted).

Reprinted from Out of the Crisis *by W. Edwards Deming by permission of MIT and W. Edwards Deming. Published by MIT, Center for Advanced Engineering Study, Cambridge, MA 02139. Copyright 1986 by W. Edwards Deming.*

The result of the approach was that consumers have come to expect and live with a level of quality dictated by cost-benefit analysis. Automation systems, for example in auto manufacturing, were designed to produce goods based on an acceptable level of defect philosophy. While consumers bought American, there was little choice. One simply accepted the level of quality proffered. But, the rise of Japan as an economic power after World War II coupled with the Japanese philosophy of satisfying customers with quality began to offer American consumers choices they had not previously had. Again, the auto industry provides the perfect example. Thus, to the extent that the Japanese and American approaches produce products and services with different levels of consumer satisfaction as a whole, the conflict will be between management approaches, not individual products. In other words, present-day American corporate culture cannot survive the force of new foreign competitive forces.

But the issue is not whether western management will be transformed by this "philosophy" but that it has been *influenced* by it. And that it has. Recognition is growing that American corporations must have an organizational culture that is suitably adapted to the new exigencies in international competition. What are these exigencies? A partial list might include the following:

- Emphasis on customer satisfaction
- Improved quality
- Greater flexibility and responsiveness to changing market conditions
- Innovation and creativity

Emphasis on Customer Satisfaction

Arthur Taylor argues that American managers have lost sight of certain fundamentals related to "franchise maintenance."

> A company that has a product or a service that is not pulled through the market place by the public, by the consumer, does not have a franchise at all. A franchise that requires pushing the product or the service through by excessive promotion and advertising expense is a fine franchise until your international competitors arrive in your market. Very often *they* are willing to take losses over many years to buy your franchise. Very often their products have qualities that your products don't have. Very often, when that hap-

> pens, your advertising and promotion falls apart. All of this has happened to us in the recent past. (1986)

Taylor believes that much of the responsibility for the problem lies with various management approaches that have always existed in the modern American corporation. The period following World War II, for example, saw a tremendous emphasis on sales. Salesmanship became "high art" and those who practiced it well were promoted rapidly. But in the late 1950s and 1960s, salesmanship gave way to the financial experts. As operations became more complex, a trend accelerated by the rise of the conglomerateur, the use of "sophisticated financial controls" became more prevalent. Taylor argues that it was the blind faith many put in the system of financial controls that left American managers "blind-sided" to many aspects of international competition.

> We will never know how many good investments were thrown away, how many fine products were never manufactured and how many tens of thousands of employees were misdirected in the name of return on equity. (Taylor, 1986)

In recent years, dissatisfaction with our system of financial controls has been increasing. Taylor and others believe that with pressures from international competition and the force of technology in its sails, marketing as a vehicle for satisfying customer demand will come to dominate corporate decision making. Some evidence suggests that American managers see information technologies as having a role maintaining close customer contact. In the Fordham survey, for example, almost 90 percent of the respondents said they were using information technology as a means of strengthening their relationships with customers, whereas only two-thirds said the same was true of relationship with suppliers.

Improving the visibility and influence of marketing as a response to pressures in international competition is, of course, an organizational policy in its own right. But, the implications for marketing technology are twofold. First, information technologies are beginning to play an increasingly major role in *marketing* by creating new organizational forms, new products and services, and enhancing existing products and services. Of special note is the fact that electronic marketing typically affords the opportunity of gathering much more information about the consumer in the act of selling.

At the simplest level, computers have helped create an explosion in "direct marketing," activities where prospects are pinpointed individually. The act of using a credit card for a purchase gives the seller access to name and address (a mailing list) and, through connections with other consumer data bases, information about the history, credit-worthiness, and preferences of the owner. At the opposite end of the spectrum from the emphasis on the individual, a far more complex form of consumer data collection is emerging. Information is collected daily from large populations in dozens of major markets covering a broad range of products and services. Supermarkets compile purchase profiles of specific customers whose identity is coded into the computer. Test marketing of new products is being increasingly automated. John D. C. Little of MIT believes this will eventually produce a syndicated market response analysis system in which

> . . . data suppliers, and probably not the manufacturers themselves (except a few big ones) will analyze the data for all competitors in coffee, detergents, cereals, or what have you. They will do it automatically and will produce reports on the effectiveness of marketing variables, publishing them in electronic media and hard copy. (1986, p. 33)

Thus, information technologies are focusing attention on consumer needs and desires in a way never possible before. Until recently, many transactions between seller and buyer were anonymous. But now, John D. C. Little warns that we are rapidly approaching the time when a glut of data will overwhelm those it is intended to benefit and will create difficulties in analyzing and absorbing its meaning.

The second implication of the marriage between marketing and information technologies is creating the ability to respond rapidly to customer demands. As we have already seen, information technology has the *potential* for altering economies of scale in both manufacturing and white-collar work. It also has the *potential* to improve quality and increase variability in products and services. Thus, coupled with the attitudes of our international competitors that quality, productivity, and customer satisfaction are supreme considerations, considerable pressure exists to ensure that additions to the firm's technological infrastructure are in harmony with those attitudes.

Technologies that are based on large-scale economics cause businesses to distance themselves from the customer in two ways. First, because products and services are homogenized, customers have to give up certain unique characteristics. But, secondly, this has led many American businesses to exert considerable pressure to shape consumption patterns. One only need consider the automobile industry's insistence on building large, energy-consuming vehicles when American buyers were switching to smaller, energy-efficient cars. This was motivated to a large extent by large-scale production economics; without sufficient and predictable demand, major investments in capital equipment are risky and difficult to justify. New distributed processing technologies—office automation and computer-integrated manufacturing—are removing large-scale economic constraints. Essentially, newer technologies permit profitable small-scale production and exploitation of smaller, specialized market niches.

Improved Quality

Quality has become an important issue in business for two reasons. First, the availability of quality products and services from foreign competitors at a low cost helps set the level of expectation among consumers. With increasing emphasis on customer satisfaction, quality is enjoying a resurgence as a primary competitive factor. The second reason relates to improving the quality of the process. As we have already seen, the economics of computer-integrated manufacturing depend in part on reducing variability and keeping all components continuously moving in the system. When parts or materials are on the floor or waiting to be reworked, much of the benefit is lost.

Disciples from the Deming school of thought, among others, argue that quality should be achieved through continuously improving the process, an approach which has come to be known by some as *integrated process management*. According to Deming, management should:

> improve constantly and forever the system of production and service, to improve quality and productivity, and thus constantly decrease costs. (1986, p. 23).

and:

> create constancy of purpose toward improvement of product and service, with the aim to become competitive and to stay in business, and to provide jobs. (1986, p. 23).

The "points" as Deming calls them, apply to the service industry as well as manufacturing. This approach eliminates the need for mass inspections and quotas. By constantly striving for zero defects, the relevance of these practices becomes moot. In the United States quotas are common. A mass transit system, for example, might set a quota of operating only 10 percent or less of its buses with defective air conditioners during the summer months. But this quota is meaningless because the transit authority should be continuously *striving* for zero percent broken equipment.

Information technologies offer an almost unique opportunity to implement quality-related policy. As technology infrastructures play a larger role in both factory and office, they will be a major determinant of quality in products and services. If quality is built into the systems from the beginning, that quality will be reflected in all results produced by the infrastructure. A *basic characteristic* of a technology system is that once constructed, it continues producing indefinitely at the level of quality it was designed to produce. *Technology institutionalizes quality.* Efforts to continuously improve the technology system are also likely to pay handsomely because as each improvement is made, it too becomes a part of the overall system. Of course, this effect works in reverse as well. Systems can atrophy if neglected. As time passes, they become less relevant to the business, which tends to evolve to take advantage of new or different opportunities. As systems are modified and patched, they can become less efficient, less quality-oriented.

In the past, information systems have typically been managed on a project-by-project basis using systems analysis and design techniques to develop applications in relative isolation. Once an application was developed and implemented, it was rarely looked at again except when maintenance was needed to correct problems or when the system was clearly outmoded and needed modification or replacement. Technical staffs were structured around these objectives: project development and trouble-shooting. Some *minor* process improvement may have occurred, but very little.

If quality through continuous improvement of the technology infrastructure were stated corporate policy, then the question would be: How is the policy to be implemented? One school of thought, typical of American management style, is that a separate unit within the technical function should be created to address quality-related concerns. After all, if the job is to get done, someone must have the responsibility. This approach turns out to be defeatist, however. By creating a quality control unit one in effect says that quality is a special job function. Routine project-development and maintenance people regard quality as "not my job." Thus, the quality unit and other units tend to neutralize each other. If a quality unit were the answer, then the solution to the problem would be *structural.* Dedication and understanding of quality must be part of the design and ongoing improvement of the technology infrastructure.

How managers change American corporate culture to be quality oriented is dependent on the creativity and originality they use to tackle the problem. Some have tried to inculcate quality through techniques used by the Japanese: quality control circles, employee suggestion progams, and the like. But, one program at IBM's Lexington plant demonstrates how quality in production can be as American in concept as apple pie and motherhood. When IBM converted its plant to computer-integrated manufacturing, remaining employees were reassigned to carry out maintenance of new equipment. One employee was assigned to each block or "cell" of equipment, and these people were designated "owner/managers." The sense of personal responsibility communicated by this designation is, of course, reminiscent of a time when the individual entrepreneur carried on most production, in a less automated day and age. But even with highly automated manufacturing, IBM was able to create a corporate culture based on quality.

Flexibility and Responsiveness

Someone recently told me of a Tawainese company that was so sensitive to market demand that it had discontinued producing women's dresses and begun manufacturing electronic components. While this reaction is extreme, it does demonstrate the range of variability American companies now face from some international competitors. Not all companies face competition where such rapid turnover is the norm but evidence suggests that an increasing number do. The financial services industry, one example frequently cited throughout the book, is evolving into a rapidly changing, consumer-oriented marketplace.

As with quality, the trend towards flexibility and responsiveness comes from two major sources: expectations created by foreign competitors and expectations created by technology's greater capabilities. Specifically, information technologies have acted to decrease *cycle time* for many businesses—the time that it takes to design, engineer, produce, sell, and deliver a new product or service. In publishing, for example, it takes an average of eighteen months from the time an author submits a manuscript to actual publication. Electronic editing, such as the type used by McGraw-Hill (see Chapter

6) reduces this time drastically. The result is that topical books get to the market in a much more timely fashion.

Ford Motor Company and other major automobile manufacturers now use computer-assisted design to create new body styles which then can be "road tested" without building a mock-up. Companies designing computer-integrated manufacturing systems can simulate operations without building a physical prototype. These uses of technology and countless others are chewing away at the time it takes a product or service to go from conception to the marketplace. Again, we see the same theme emerging in regard to responsiveness and flexibility as with customer satisfaction and quality. *Information technologies can accommodate these corporate cultural values;* the issue is whether or not *organizational policy* is consistent in its support.

Innovation and Creativity

Creating an innovative corporate environment is important for the timely introduction of technology. The policy issue for management is to create a climate for innovation that opens the organization up to the absorption of new technologies as rapidly as needed. According to James Brian Quinn:

> Innovation—creating and introducing original solutions for new or already identified needs—must be one of the central themes for society and for technological management during the next few decades. (1979, p. 19)

Evidence suggests that innovation is more important to managers in companies overseas than to U.S. managers and figures more centrally into their approach to competition. Arthur D. Little, Inc., the Cambridge, Massachusetts-based consulting firm, conducted a three-year study of senior executives in the United States, Canada, Europe, and Japan that documented these differences. The study found, for example, that two-thirds of the Japanese managers said that the need for innovation in their organization was very important, compared to 41 percent of the North American (U.S. and Canadian) managers. When asked whether they thought that innovation would make *specific* contributions to earning in the next five years; 87 percent of the Japanese managers, 71 percent of the European managers, and only 51 percent of the North Americans said "yes."

Most managers believe that innovation can be managed. Two-thirds to three-quarters of the managers surveyed by Arthur D. Little, Inc., regardless of nationality, said that innovation can be managed but specific skills and knowledge are required to manage it well. The remainder believed either that innovation can be managed in much the same manner as other corporate functions or that it cannot be managed in the conventional sense of management. In any event, the great majority of managers in all countries believe that the time they will spend on management of innovation is increasing.

Innovation shares a close kinship with quality, flexibility, and responsiveness because of its association with satisfying customers. Large companies that are successful innovators seem to share certain common characteristics which include an encouraging climate for innovation, long-term vision, and a focus on small teams and interactive processes.

Climate and Vision

Senior managers create a climate that encourages innovation. The corporate culture tends to be dominated by a long-term vision which is not expressed only in quantitative terms. Management support is explicit, providing information to shape and develop ideas, support, and resources to carry it through. Yet no matter how long-term the vision, both the technology infrastructure and innovation should be oriented towards serving the marketplace. As we have already seen, emphasis on customer satisfaction is an important exigency in the international marketplace.

Small Team Focus

Much innovation takes place in small, highly-committed teams sometimes referred to as "skunkworks." The objective in establishing such teams is to insulate the innovative group from bureaucratic influences and encourage a free, interactive exchange of ideas. Individuals who participate in innovation must, at one time or another, serve in one or more of the following roles (Roberts and Fusfeld, 1981):

Idea generator—analyzes and synthesizes information from markets and technologies to come up with ideas for new products and services

Champion—acts as a sponsor and promoter of the innovation, and can be technical or management

Project leader—handles planning and coordinating activities

Gatekeeper—transfers information into the organization

Coach—advises and guides less experienced personnel, a mentor

People who work well in an innovative, entrepreneurial environment tend to have personality traits normally unsuited for organizational life on the whole. These traits include difficulty in working with authority. A focus on a small group can, therefore, have the effect of buffering individuals from these influences.

Interactive processes, multiple approaches

Often innovation is best carried out by setting up several task groups to work on competing approaches. At some point, management must commit to further development of a single idea and abandon all others. Thus, redundancy and risk in the form of unsuccessful and incomplete approaches are a *normal* part of the process and should not necessarily be considered wasteful. This viewpoint is considerably different from the bureaucratic perspective that commits a single team to the task of solving a problem and depends on a successful outcome.

Though the bureaucratic form of organization endures, the next few decades undoubtedly will require that it be sensitized to the need for innovation. An innovative climate is companion to rapid introduction of information technology and a conscious policy effort should be made to break down barriers. The problem is typically more acute in larger organizations. There, size and complexity have caused bureaucratic mechanisms to be exaggerated in the direction of structure, standard procedure, planned time horizons, and other forms of control. Leaving organizational resistance intact while attempting to implement new technologies is akin to attempting to sow seeds without first tilling the soil. The resulting crop is likely to be largely happenstance.

Technology and Structure

Just as international competition has teamed with information technologies to create pressures on organizational culture, these forces are also likely to cause changes in structure. Before exploring what those changes might be, two caveats are in order. First, changes in structure will occur within the context of bureaucracy and hierarchy; they will not replace them. Examples were noted earlier in the chapter: hierarchy serves an important role in organizing communication patterns and permits task specialization. Second, structure varies from company to company and for good reason. Structure is a very special element of an organization's response to the tasks confronting it. Different tasks require different organizational structures; and as one might expect, technology infrastructure should match or serve structure, not the other way around.

The second point requires some elaboration. Charles Perrow (1986) demonstated some differences by analyzing companies on two dimensions: coupling and interaction. Coupling refers to how closely knit different components of the organization are. Tight coupling permits rapid decision making, strict schedules, and quick responses to changes and deviations within the organization's systems. Loose coupling, on the other hand, provides time for members of the organization to devise alternate strategies, substitute personnel and equipment, and implement other methods of containing errors and systems failures. For example, bureaucracy with its slow communications and formal procedures can sometimes prevent overreaction or inappropriate response to some external stimulus. Universities, many government agencies, and traditional manufacturing tend to be loosely coupled while airlines, consumer electronics firms, and news media are relatively tightly coupled.

The second of Perrow's comparisons is the nature of the interactions, which can be either linear or complex. In linear interactions, processes tend to be routine, well-understood. Few unfamiliar or unintended circumstances occur. Assembly-line production and railway transportation are good examples of linear systems. Complex interactions include unforeseeable and unexpected circumstances. Normally, interactions occur among multiple components as is the case with airlines and broadcast television.

By combining the two dimensions as shown in Figure 7–2, Perrow demonstrated how a company differs markedly in terms of its needs for centralization or decentralization. Cell 1 depicts companies that are tightly coupled and have linear interactions. A centralized structure is needed for rapid response and decision making characteristic of tight coupling. Centralized structure is also compatible with linear interactions. The types of organization one would expect to fit this mold include dams, power grids, some continuous processing, rail and marine transportation. One would also expect the technology infrastructure to match; data processing is probably *least* distributed (or decentralized) for organizations in this cell.

Cell 2 changes the combination by introducing complex interac-

		Interactions	
		Linear	Complex
Coupling	Tight	Cell 1 *Centralization* for tight coupling *Centralization* compatible with linear interactions (expected, visible)— dams, power grids, some continuous processing, rail and marine transport	Cell 2 *Centralization* to cope with tight coupling (unquestioned obedience, immediate response) *Decentralization* to cope with unplanned interactions—nuclear plants, space missions
	Loose	Cell 3 *Centralization* or *decentralization* possible (few complex interactions); tastes of the elite and tradition determine structure— most manufacturing, single-goal agencies (motor vehicles, post office)	Cell 4 *Decentralization* for complex interactions desirable *Decentralization* for coupling desirable R&D firms, multigoal agencies, universities

Figure 7–2. Interaction/Coupling Chart.
From Complex Organizations: A Critical Essay *by Charles Perrow. Copyright © 1986 Random House, Inc.*

tions. Organizations that fit this category include nuclear plants, chemical plants, and space missions. Centralization is still needed for tight coupling but some decentralization is expected to cope with complex interactions. The technology infrastructure would follow suit with a strong central data processing unit for some tasks and distributed processing for other tasks. Cell 3 combines loose coupling and linear interaction. Perrow indicated that the degree of centralization or decentralization could vary depending on management's preferences. Again, the technology system could vary as well, but the form would follow the structure. As an aside, Cell 3 also demonstrates how technology systems can be designed in different ways in response to a single organizational type. The variability in design will depend on differences in structure, which can also vary.

Finally, Cell 4 depicts the totally decentralized situation: loose coupling and complex interactions. Research and development firms, multigoal agencies, and universities are representative organizations. All tend to have diverse, complex technology infrastructures.

What does Perrow's model teach us? It teaches that while information technologies might create the *potential* for structural change, that change might not be desirable. We should also keep in mind that a company can be a multifaceted organization with a number of different task sets. Thus, one part of the organization may be very different from another, requiring separate structural and technology responses.

Structural Evolution

Information technologies and structure have a mutual influence on one another. Chapter 3 traced the evolution of technology from central data processing into the highly fractionated, diverse end-user-oriented environment. Structure also moved from being centralized and functional to a decentralized, multidivisional approach. Divisions became important strategically. And the need to integrate certain activities while specializing gave rise to the matrix form of organization. Some companies have used task groups in addition to the underlying matrix to address special strategic tasks. But some experts believe that the organization will evolve further to become amorphous. Rather than having a fixed shape, the organization will be temporary and fluid, restructuring itself as needed to address contemporary contingencies.

Raymond Miles and Charles Snow (1986) describe this new organizational form as a "dynamic network." Its chief feature is that it vertically disaggregates the traditional organization (see Figure 7–3). Independent companies handle design, production, material and labor supply, and distribution. McGraw-Hill, a company previously cited, utilizes a network for functions it previously performed itself while concentrating on product development. Entrepreneurial brokers in retail clothing are combining American marketing and distribution concerns into networks with overseas low-cost production to produce a dazzling array of new labels. The dynamic network approach does offer a company several advantages. It permits efforts to be concentrated in an area of distinctive competence. It permits flexibility and responsiveness to rapidly changing market conditions. It is essentially an opportunistic form of organization. Companies organized as dynamic networks also will not be saddled by aging, less

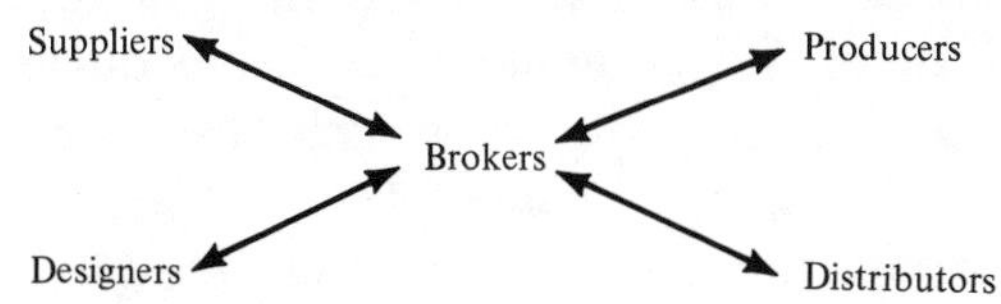

Figure 7–3. The Dynamic Network.

From "Organizations: New Concepts for New Forms" by Raymond E. Miles and Charles C. Snow. © 1986 by the Regents of the University of California. Reprinted from the California Management Review, *Vol. 28, No. 3, p. 65. By permission of the Regents.*

efficient equipment. As a producer loses cost competitiveness, customers (other networked firms) will seek out more efficient producers.

Networked organizations are most likely a healthy addition to the variations on organizational form because their fluidity more easily accommodates innovation and creativity. However, networked organizations are likely to play a limited role, based on need for this particular form. They are, almost by definition, loosely coupled and interactively complex. We must not overlook the fact that in some industries—financial services and travel, for example—a recent resurgence of integration has been occurring. Banking, investments, and insurance are molding into integrated financial services. In airlines, major carriers and regionals are merging into large integrated systems like Texas Air. And attempts to create companies like Allegis are demonstrating how they are becoming integrated travel services combining airlines, hotels, auto rental, and travel agents. One should remember that the reintegration trend in airlines was preceded by a period of rapid growth in small specialized regional and low-cost carriers. This in a real sense represented the disaggregation phase of that industry.

What all this means is that the dynamic network may be a transitional form of organization. An industry that must restructure itself for competitive reasons first disaggregates, then reorganizes relationships into a more productive pattern, than reintegrates. This follows the familiar "unfreeze-movement-refreeze" organizational change model. Airlines disaggregated immediately following deregulation, were reorganized through a series of mergers and acquisitions, and are now reintegrating into a new industry structure. Financial services have followed a similar cycle. We now see other industries such as manufacturing currently in the disaggregation phase. But because manufacturing is not regulated as airlines and financial services were, the forces driving disaggregation are different. In order to seek advantages among network components, firms must typically move production overseas, often simply contracting it to foreign firms. The disaggregation of American manufacturing concerns means that at some point reintegration will occur, as it already has begun to do in the automobile industry. Reintegration is likely to occur in the form of the corporation that is truly international: its ownership, creditors, and loyalty will all be strictly international.

The services sector is unlikely to be spared this experience as well. Already one eastern-seaboard company ships software programming tasks to an office in Ireland where labor is plentiful at reasonable cost. Once coded, programs are transmitted back to the United States via telecommunications. With the falling cost of telecommunications, such arrangements are likely to become more common. New York investment banking concerns which are moving "back office" operations from Manhattan to the less expensive environs of New Jersey may someday by moving these offices to India, China, South Korea, or Mexico.

Enormous implications flow from this observation. The headquarters of the eastern-seaboard company that sends programming tasks to Ireland happens to be within blocks of a low-income neighborhood with high unemployment. Unemployment among Americans is a major concern that will be considered in Chapter 8. These structural changes also raise serious questions about how the United States will compete on a global basis, an issue taken up in Chapter 9. Finally, there are concerns about how management should think about the technology infrastructure. In industries that are on the threshold of disaggregation only to reintegrate at some future point the question becomes: What general competitive capacity is needed from information technologies? Clearly, the folks at People Express did not understand this issue for, as we have already seen, their information system was designed without the ability to handle complex fare arrangements. If networked organizations are primarily transitional or occupy a specialized niche, then we must ask what will happen to *large organizations*? Part of the answer, as already noted, is that hierarchy and bureaucracy will endure, although they certainly will be sensitized to needs for innovation and responsiveness. The familiar organizational pyramid will go through some contortions as well and, in fact, already has. Richard L. Nolan and Alex J. Pollock (1986) contrasted organizations of 1960 to those in 1985 (see Figure 7–4) and concluded that the familiar hierarchical pyramid has evolved into a networked diamond-shaped structure. This conclusion is based on the distribution of workers at various levels. In 1960, for example, top management constituted 5 percent of the total structure, middle management 35 percent, and operational and clerical employees the remaining 60 percent. By 1985, top management still represented 5 percent, middle level employees (including middle management and a new class of knowledge-workers) 55 percent, and operational and clerical workers 40 percent.

But, how has the diamond-shaped organization come about? Many experts are predicting and evidence suggests that middle man-

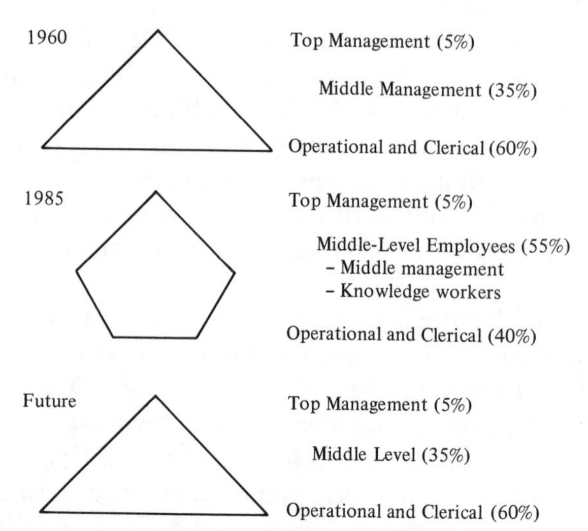

Figure 7–4. Evolution of Large Organizations.
Adapted from "Organization and Architecture, or Architecture and Organization" by Richard L. Nolan and Alex J. Pollock. Stage by Stage *(September–October 1986), p. 4.*

agement layers are being substantially reduced as a consequence of white-collar automation. Here, the middle level bulges. The answer is evolutionary. Information technologies have had their greatest impact on lower-level operational and clerical employees. Thus, this level has been reduced disproportionately more when compared to the whole. Automation of managers and professionals has only begun and one would therefore expect this level to decrease at a much greater rate over the next few years either top management or operational and clerical levels. However, as the middle level is automated it will not disappear as many have claimed. Information systems specialists are one category of middle-level employees, for example, that will stay. Many of the other functions performed by this group will remain. Indeed, middle-level employees continue to serve as organizers of communications in the hierarchy. In other words, the large organization is likely to return to the tried-and-true hierarchical pyramid shape at some point in the future. The pyramid's productivity will be higher. As Nolan and Pollock point out:

> For many companies, the total size of the workforce has remained relatively constant over the last decade, while sales revenue and unit volumes have increased significantly; in some aggressive companies, unit volumes have more than doubled. (1986, p. 4).

Technology and Management Style

We have already seen that results from implementation of technology are likely to be disappointing, even nonexistant, when organization culture and structures are not consistent with the technological infrastructure and business goals. Management style is essentially a policy tool for communicating or promulgating the appropriate climate throughout the organization. Change is stressful enough for organizations and is often a most difficult task to accomplish. Mergers involving two distinctly different organizational cultures have been known to fail because of the difficulties of combining them. And even when cultures are married, years must normally pass before the "separatists" are no longer a factor in the organization. General Motors' acquisition of Electronic Data Systems provides one example.

The introduction of new technology, especially on a massive scale, is little more than an attempt to combine different cultures. It is not uncommon to find two distinct camps (technologists and business managers) around which debates swirl about appropriate management style. Traditional managers argue that the technologists do not understand the business; technologists argue that technology mandates a completely different management approach to which traditional managers cannot relate. The pros and cons of these arguments are considered in Chapter 8. Thus, the challenge of combining the traditional and technology is difficult enough.

Management style occupies a central role in organizational change. Waterman, Peters, and Philips point out that "a corporation's style, as a reflection of its culture, has more to do with its ability to change organization or performance than is generally recognized." (1980, p. 358). Style takes on a symbolic quality that reinforces management policy. Style backs words up with action and makes the culture's message more believable throughout the organization. This is why reports that American managers give only lip service to technological and organizational change are so disturbing. Respondents to the Fordham survey, for example, reported that their greatest frustration in managing technology was senior management's failure to make good on their stated policies and do more than give them nominal support. Providing lip service or simply sanctioning or "lending support" to technology is not enough. It misses the point. Effectively managing technology *starts* at the highest levels with management setting the direction of the company. Appropriate modifications to organizational culture and structure issue from this direction, as do needs to construct or modify the technology infrastructure. All should be reinforced with appropriate management style.

Many issues of management style are related to the workforce, and these are taken up in Chapter 8. However, two issues related to the organization as a whole are examined in more detail here: inertia within the organization and the importance of creativity.

Overcoming Inertia Within the Organization

The spector of technology-related innovation almost invariably provokes resistance within organizations. Left unattended, resistance has been known to cause the failure of some information system projects. It can occur throughout the organization, from the highest levels all way down. Resistance is largely an emotional issue. It can and should be managed.

According to John P. Kotter and Leonard A. Schlesinger (1979), there are four common reasons why people resist change. First, they are afraid they will lose something of value. Often what people regard as being of value may be more psychological than tangible, an implicit "contract with the organization." Individuals and groups may display overt or subtle political behavior in protecting what they perceive to be their best interests.

Ironically, one antidote for protection of self-interest may be *creating conflict* between individuals and groups. Constructive conflict can be an important mechanism for shaking up the organization and getting its vital juices going. Used properly, conflict can bring out competitive spirit and incentive. It can heighten morale and arouse people to solve a problem that might otherwise go unattended. The "straw dog" approach whereby an unacceptable solution to a problem is presented as a fait accompli is one example. People affected by the "solution" react by proposing a more effective solution of their own making. In this way, conflict stimulates organizations to think through ideas and avoid uniformity in thinking. But in thinking through, people forge into new territory and create new rules, new traditions, and new values. Conflict can integrate.

According to Lewis Coser:

> No group can be entirely harmonious, for it would then be deviod of process and structure. Groups require disharmony as well as association; and conflicts within them are by no means altogether

> disruptive factors. Group formation is the result of both types of processes. The belief that one process tears down what the other builds up, so that what finally remains is the result of subtracting the one from the other, is based on a misconception. On the contrary, both "positive" and "negative" factors build group relations. Conflict as well as co-operation has social functions. Far from being necessarily dysfunctional, a certain degree of conflict is an essential element in group formation and the persistence of group life. (1956, p. 31)

Conflict is not always constructive, however. If induced, it must be fair, with all groups having a stake in a satisfactory outcome that furthers the objectives of the organization. And conflict is sometimes difficult for American managers to accept due to cultural biases. Most managers view conflict as something to be weeded out and destroyed, not managed.

The second reason people resist change is that they do not understand it or its implications. Especially in organizations that lack a high level of trust between employees and management, people misinterpret the motives behind the change. Technology is often seen as a way not only to displace workers but to redesign work itself in favor of management. Ipso facto, it is to be resisted. The increased emphasis in Japanese and European and some American companies on teamwork between employees and managers is to some degree a logical outcome of increased technological innovation within these firms. For as more technology is introduced, the need to flush out misunderstandings and rapidly clarify them also grows.

Third, people often believe that the change is not in the best interest of the organization. As noted earlier in the book, no single correct technological approach is appropriate. Indeed, with Perrow's loosely coupled, linear organizations, use of centralized or decentralized structure depends on management's preferences and tradition. Information technologies have various "camps." Some believe that small personal computers are an evil visitation and that large, centralized machines can handle all the organization's needs. Others believe the mainframe computer is obsolete (or should be). Even different vendors have camp followers. A company is described as being "true blue" (all IBM machines) and so forth. Many of these differences stem from the fact that people are basing viewpoints on incomplete information. Even with complete information, multiple technological approaches—all valid—are possible. Yet the roads not taken are likely to be populated with malcontents who resist.

Fourth, there are, of course, people who have a low tolerance for change. People who are older, who have less education and training, who are lower in the hierarchy, and who have fewer career options are likely to be more change resistant. This resistance is not always negative. Many senior managers occupy their positions because they demonstrate maturity and caution in their decision making, and are not easily swayed. If all resistance were somehow removed, organizations would disintegrate because nothing would endure. An important role of management and the organization itself is to provide continuity. When resistance translates itself into a mechanism for ensuring that change is constructive, it is valuable. A basic premise of this book is that management has long been advised that technological change creates a need to change ways of managing. Certainly technology poses new challenges and new problems, but the basic roles and responsibilities of management remain. The notion that innovative management practices are needed has, in itself, set in motion a dynamic of resistance that ultimately translates into lost momentum with technology. Lost momentum with technology translates into lost competitive advantage for a firm within the industry or for the United States within the global community. Deming (1982) said: "Drive out fear, so that *everyone* (emphasis added) may work effectively for the company."

Various strategies for dealing with change are available. Although an extensive treatment is beyond the scope of this book, a few are briefly described below.

Communication

Resistance can often be substantially reduced by simply communicating the reasons for and importance of technological change. Even in cases where the change has negative effects, understanding the issues can help. Labor unions have often cooperated with conversions of manufacturing plants because they understood the choices faced by management: losing jobs to automation or the entire business to overseas competitors. Communication may also play a much more important role in rallying support for technological innovation in a different way. Changes and the abstract quality of work with computers increasingly cause employees to feel out of touch with what work is about. In contrast to the craftsman or even factory worker of decades ago, today's worker may be distinctly less clear about the contribution he or she is making. This situation is not helped by the restless restructuring of corporate interests through mergers and acquisitions.

Ironically, information technology itself may be a critical component of the solution to the communications problem. According to Shoshana Zuboff, information technology not only automates operations, replacing human effort and skill with technology to reduce cost and provide greater control and continuity, but "informates" as well. "Informate" is a word coined by Zuboff which means that automating an application will simultaneously generate information about the organization's underlying work processes. For example, informating occurs in manufacturing when computers capture data from robots and sensors about production processes and display the results on a video terminal or computer printout. And:

> In the office environment the combination of on-line transaction systems and communications systems create a vast information presence that includes many data formerly lodged in people's heads, in face-to-face conversations, in discrete file drawers, and on various pieces of paper widely dispersed in time and space. In its capacity to automate, information technology has a prodigious ability to displace human effort and to substitute for much that has been familiar as human skill. As an information technology, its implications are equally significant, although not yet well understood. (1985, p. 8)

Education

Education issues related to technological innovation are complex and encompassing. Education has so many roles, one of which is enhancing communication by providing organizational members with a common language. Managers in one company long ago recognized this interrelationship between education and communication. All employees, regardless of job, were trained in the language of finance. Then, management used a low-level technology—grease pencils and a projection system—to display current, key financial results to production workers. The message was clear; workers could understand for themselves where the company stood.

Most businesses lack a common language, and technology will offer an important medium for communicating throughout the organization along with the language of finance, accounting, and marketing. Most technology-related education today is driven by *immediate* need: job skill training. But breaking down resistance means equipping all members of the organization with the basic language of technology to help resolve misunderstandings and permit an informed, direct assessment of events.

COMMITMENT AND CO-OPTATION

Another method of dealing with resistance is giving potential resisters a role in the change. Kotter and Schlesinger (1979) point out that participation generally leads to commitment, though not always. Participation may be specially significant in cases where potential resisters have knowledge or information relevant to the introduction of new technology. This is increasingly the case in recent years as the knowledge-worker has become the focus of automation efforts. Consider the feelings of an engineer or investment portfolio analyst who is considered an outstanding expert in his or her respective field, even if that reputation is localized within the company. Capturing that knowledge with expert systems (artificial intelligence) technology and blending it with other experts' views, which is frequently done when developing expert systems, robs personal identity. It belittles the lifetime career of the individual spent accumulating, refining, and perhaps even expanding the knowledge.

In cases like this, resistance can be reduced by giving the engineer or manager a key role or at least participation in the process of introducing the technology. Experience has verified this observation. Many experts have become committed participants of projects that make use of their special skills. But, participation and co-optation have serious downsides; unless properly managed so that participation leads to commitment, one is left with a resister within the ranks of the project.

NEGOTIATION AND COERCION

Negotiation can be used to deal with resistance in cases where one party is likely to lose as a result. Again, technology's displacement of labor, the example used previously, comes readily to mind. Negotiation might be formal or informal; an example of the latter case would be offering incentives to experiment with new technology (e.g., flexible hours, promotion). Coercion involves forcing resisters to accept changes and is successful primarily in situations where the would-be resisters have little power.

The Rising Importance of Creativity

Over the next decade, the number of knowledge-work jobs will increase 50 percent faster than the growth of the labor force. Jobs in service industries are growing three times faster than in other sectors. Professional and technical jobs as a category are growing at the fastest clip of all. Despite the rising demand for knowledge-workers, the pool of available people is actually projected to decrease. Clearly, by the 1990s competition for qualified people will stiffen. All this comes at a time when many businesses are complaining that the education and skill level of new entrants to the labor force is declining and educational institutions are falling short of their mission. These problems come at a time when the stakes in knowledge work seem to be rising. The growing emphasis on knowledge work is pushing the workforce as a whole towards a different perspective. Knowledge-workers are more tolerant of ambiguity and uncertainty. They accept change more readily. They are characteristically problem solvers and are more *creative* and imaginative in seeking solutions.

The trend towards a more creative workforce is reinforced by technology. Technology becomes a way to extend their energies by automating knowledge and moving on. In Chapter 3 we observed how expert systems technology is beginning to capture knowledge. The pipeline theory (Chapter 5) told us that once knowledge is effectively captured, the focus of organizational activities changes to leveraging knowledge or *creating new knowledge,* and ability to create new knowledge is likely to be a distinguishing factor among competing organizations. The act of creation, or creativity, is a special subcategory of innovation. Technological innovation within organizations, at least the way we have used the term here, may involve merely new applications of technology or even routine technological development that is not dependent on creativity. Creativity involves producing something that is novel and, if it is to ultimately contribute to the organization, relevant.

What does this mean for management policy, the technology infrastructure, and for business competitiveness? Clearly, as knowledge becomes more important over the next decade, organizational and technological resources must be committed to its preservation and creation.

Conclusion

This chapter is based on three major assumptions about information technologies and organizations. First, there is growing pressure from foreign competition and an accelerated real need for corporations to reorient organizational processes. They must generally be more responsive and flexible. Satisfying customers through quality is a central focus, and organizations must be less resistant to change and more adept at leveraging their know-how (i.e., knowledge). Positive actions by managers are needed to make appropriate adjustments in the organizational culture.

Second, organizational changes that do occur should be tied to the competitive circumstances faced by the company. Thus, business direction should drive organization, and organization should drive the form and content of the technology infrastructure. Because technology *permits* certain organizational changes to occur does not mean they should happen.

Finally, no matter how transformed the organization, it will continue to function as a hierarchy with bureaucratic characteristics. Management will accomplish the desired changes within this time-honored framework. What this suggests is that radical management approaches are not appropriate. Most of the knowledge and skills needed to make the changes can be acquired, or already exist, by even the most ardent curmudgeons in senior management.

8

Technology and the Workforce

The Social Context

An important theme in the Integrated Technology Management Framework is that changes in the social culture of the organization are at least as important as accurate assessment of technology and decisions about investments in technology. We have already seen that technology infrastructure should be designed to "follow" organizational form, which in turn should be appropriate to the organization's mission. Organizational change that is accompanied by technological change will inevitably impact workers at all levels and it follows that management must, as part of their overall efforts to implement technology, consider the implications.

It is impossible to say at the outset whether the implications are negative or positive. Much publicity has been given to the so-called dehumanization of the workforce: computers supposedly cause physical impairments, emotional stress, and even loss of constitutional rights. Indeed, much of the research done on technology's impact on workers and society, and experience, tell us that the effects can be quite positive. Many workers *like* working with computers. The dichotomy—positive or negative—comes in large part from society's feelings about technology in a broader sense. Those feelings are divided, sometimes polarized, sometimes ambivalent. Technology means different things to different people. As David F. Noble wrote:

> Technology has been feared as a threat to pastoral innocence and extolled as the core of republican virtue. It has been assailed as the harbringer of unemployment and social disintegration, and touted as the creator of jobs and the key to prosperity and social stability. It has been condemned as the cause of environmental decay, yet heralded as the only guarantor of ecological integrity. It has been denounced as the handmaiden of exploitation and tyranny, and championed as the vehicle of emancipation and greater democracy. It has been targeted as the silent cause of war, and acclaimed as the preserver of peace. And it has been reviled as the modern enslaver of mankind, and worshipped as the supreme expression of mankind's freedom and power. (1984, p. xii)

Thus, one of the factors that makes information technologies difficult to manage is that it is a political, emotional, and value-laden issue for many people. It is political for two reasons. First, as already noted, it is a powerful engine of economic growth and, as such, is important to the ability of the United States to remain competitive in the international arena. Indeed, because of the comparatively high cost of American labor, technology has become a mandatory component of production in some industries if they are to remain price competitive with foreign companies which enjoy vastly lower labor costs. Without success in using technologies to cut the cost of American manufacture or service production, either jobs go overseas or the American real wage must decline. In essence, successful application of technology has come to mean to the American economy what importation of raw materials means to Japan: economic survival.

But, secondly, technology can also displace workers from their livelihoods, creating social hardships for individuals and their families. The effect is doubly difficult when an entire economic region suffers loss of employment, as was the case in Detroit during the decline of the American automobile industry. Workers are also voters, of course, which often brings government into the situation. If the issue were as simple as a group of workers losing their jobs because of automation, then remedies would be relatively easy to design. But the effect is much more complex than that. Technology, and especially the rapid changes in technology since World War II, have caused the economy to become much more dynamic, increasing job turnover and the likelihood that an individual will have more than one job during his or her career. Computers may also be helping to restructure the economy to a more service-oriented, lower paying, and temporary employment one. Thus, the issue for an individual worker is not simply displacement but expectations.

Technology can therefore be an emotion-charged issue. For some people, technology brings insecurities into the foreground. If it does not actually result in unemployment, it can cause a loss of personal identity. People generally derive many of their feelings of self-worth and esteem from work. When their work is eliminated or when automation of their role is proposed, they feel threatened. The threat may not actually come from the loss of job per se but from the fear of not being able to find a job that meets expectations in a world where the demand for their particular education level, age, and skill mix is declining.

But not all feelings are negative. Technology also conjures a broad range of positive human emotions. And this, as it turns out, is a key issue in managing the human-machine interaction. Many of the great advances in medical and environmental sciences, for example, would not have been possible without the computer. It would not be an overstatement to say that information technology is a key tool in stopping mankind's destruction of the planet. It has given us the wherewithal to analyze what the environmental damage is and the resulting realizations have often goaded us into action. Computers are essential elements of the drama in manned spacecraft flight; news commentators frequently refer to the role of "onboard" computers and computers at mission control. Computer graphics can also provoke a sense of wonder: The spectacular visual effects of many recent movies were created by computers that are essentially larger, more sophisticated cousins of office computers.

As computers become more human-like in their interactions with people, emotional responses are likely to become even more pronounced. Artificial intelligence technology—including expert systems, natural language understanding, and vision—will be a major contributor to this trend. Expert systems which use interactive processes to mimic reasoning and judgment-making characteristics of human experts will seem more responsive and intelligent in their dealings. Already we have seen how computers, once "taught" to be experts, can themselves become tutors to novices in the field. Natural language understanding will increasingly improve the ability of computers to understand human speech and to speak it, thus making the interface between people and machines much more natural. Vision systems with intelligence capabilities will also allow the computer to "share" certain phenomena (e.g., sight) previously the exclusive domain of humans. Touch, sound, and olfactory senses will eventually be added to the repertoire.

Just how involved people can become emotionally is demonstrated by experiences with two systems, ELIZA and DOCTOR, both developed by Joseph Weizenbaum. These systems create the illusion of natural conversation. They were developed to perform "nondirective counseling," a form of therapy in which the therapist takes a neutral role. If a human who was interacting with ELIZA, for example, said "I am tired," the machine might respond, "Tell me more about such feelings." (Frude, 1983, p. 49) The computer did not actually understand what was being said, but in most cases gave the impression that it did. Experiences with both ELIZA and DOCTOR revealed that in both experimental and clinical situations, people became emotionally involved with the systems, asking to be left in private and later claiming that the system had understood them. Many said they actually *preferred* dealing with the impartial machine rather than humans.

> Over half the clients who interacted with DOCTOR in a Massachusetts hospital insisted that they were conversing with a real person, although they had been told quite specifically that a computer was operating the terminal. Even when such animism does not occur, clients often say that they prefer interaction with the machine to that with a human therapist. Most people do not find the computer cold and impersonal, and many feel more at ease discussing highly personal topics with the machines. (Frude, 1983, p. 85)

What all this means, of course, is that people tend to take technology systems very seriously. The rationality of their feelings aside, human responses in the workplace are likely to come more and more to depend on the quality of the relationship they have with information technologies. At the simplest level, the quality of these interactions usually translates into user-friendliness and ergonomic issues. For example, how easy is the machine to use and what are the health risks involved? But more broadly conceived, they encompass the employee's entire range of attitudes, feelings, reactions, and thoughts about the workplace. It is clear that people will respond to the psychological directives of technology systems. Any management team that is heavily committed to major investments in technology infrastructure has two powerful motives for considering the implications of this fact. First, ignoring the consequences of human-machine interaction leaves the field open for potentially negative experiences. In other words, managers who do nothing will increase the likelihood that workers will have a bad experience either because they harbor negative feelings about technology or because the new technology creates a negative experience. And if human workers are unhappy with the technology, problems are likely to result. These problems are impediments to being able to derive the expected benefits from the technology.

But, secondly, information technology is a potentially persuasive medium for building a relationship between the organization and employees. Properly designed, it can help establish a beneficial corporate climate by focusing workers on key issues such as quality, customer satisfaction, or productivity. It can make workers feel good about their jobs, providing recognition, informal rewards, and other incentives.

In time, we will come to find that achieving the desired effect with technology will depend on how it is integrated with human workers. What we know now is primitive and current efforts to make use of the potential psychological bonds between humans and computers are clumsy and insensitive. For example, some companies have used the computational ability of word-processing machines to count words entered by operators to spur productivity by flashing messages on the screen such as: "The operator on your right is working faster." These cases are more a question of the shortcomings of management style than use of technology, although technology often gets the blame. But technology changes the stakes companies have in managing the psychological bond between the organization and the worker. Until now, this bond was mediated by the human supervisor for the most part. Management could set the tone, but much was lost in translation. Personal motives of human intermediaries often became an issue. But technology can by-pass the human intermediary and thus eliminate or at least considerably reduce the amount of distortion. Of course, the idea that technology could or would be used to "control" workers has raised objections among many who fear an Orwellian state of affairs: the "big brother" syndrome. However, the issue of control is academic. So long as there is technology and so long as there are people working with the technology, those people will be influenced by the technology. In that sense, technology is not nor can it be value free. The real question becomes *what* values are inherent in the systems. Often organizational values exist by default; they are the unconscious product of the failure of management to establish and promulgate positive values. They are de facto values that are renewed by the process of ongoing human interaction. But values embodied in technology are more formal and institutional. They must be explicit and articulate, not confusing. Thus, many companies that undertake major automation projects may find a need to clarify organizational values in the process of building technology systems.

People who study human-machine interactions often suggest that computer systems can and should be built to be considerate and responsive to human needs. Unfortunately, what we know about human reactions to information and control suggests that there may be no right or wrong way to design systems. Instead, managers may be faced with a series of design options, all of which have some desirable and undesirable consequences. As an example, Lawler and Rhode (1976) studied the impact of information and control systems in organizations and found that they often produce four types of dysfunctional effects:

1. *Rigid Bureaucratic Behavior.* Employees will behave in ways that make them look good in terms of the control system, but are dysfunctional with respect to the agreed upon goals of the organization. An emphasis on cost cutting, for example, can cause employees to ignore service to the customer which, in the long term, might be more important in capturing market share.
2. *Strategic Behavior.* Employees may act in certain ways in order to influence the results reported in the control system. Smoothing out earnings over several periods using various accounting techniques is one example.
3. *Invalid Data Reporting.* Simply stated, employees report data falsely.
4. *Resistance.* Employees may fail to cooperate or otherwise resist actions dictated by the control system. In Chapter 7, we discussed some of the causes of resistance.

None of these behaviors is desirable. However, when we examine a number of characteristics of information that technology-based control systems might have, we find that avoiding all negative consequences is almost impossible. We might say, for example, that data should be complete; incomplete data misleads. And, in fact, we find that incomplete data does produce bureaucratic behavior. Employees will perform in a way that reflects well in the data that is reported. But at the opposite end of the continuum, complete data tends to cause resistance among employees presumably because they feel controlled. We find the same kind of situation with objective data. Most people would agree that technology systems should report data objectively. And, sure enough, subjective data tends to cause invalid data reporting and resistance among employees. But objective data produces bureaucratic and strategic behavior. Fast communication of data often results in bureaucratic behavior, strategic behavior, and invalid data; while slow communication of data creates resistance. If employees can influence the outcome of the data by their performance, we find an association with bureaucratic behavior; if they cannot influence the data then invalid data reporting and resistance result. If data is communicated infrequently, bureaucratic behavior and invalid data reporting are the outcome, while frequent communication produces strategic behavior and resistance. Other examples could be cited.

What all this means is that the political, emotional, and value-laden technology environment presents significant challenges to management. Much of management's attitude has been "catch-as-catch-can" with respect to technology and the workforce: the problem will be solved when we encounter it. However, the complexity of the issues requires real forethought in formulating appropriate policies to accompany technology implementation. But real incentives exist for making the management effort. Those incentives include not only forestalling unintended and negative consequences where possible, but using technology as a medium to inform and motivate employees. This is far from being a simple task for two reasons. First, the employment environment with respect to technology is changing rapidly. Our understanding of how technology affects office and production workers may or may not be relevant to knowledge workers or senior managers, for example. We are always breaking ground in new territory. The second reason is that our level of sophistication in human-machine interactions is low. Although we have a broad outline of the issues that face managers where technology and the workforce are concerned, the details are sketchy at best.

What are the issues involved with technology and the workforce? They can be divided into two broad categories: (1) effect on employment levels at both the national and firm level and (2) impact on quality of worklife. The latter includes job design (i.e., how work will be performed), qualifications of employees, education, job satisfaction, and so forth. Both of these categories will be considered in more detail throughout the remainder of this chapter.

Employment Level

Several years ago, an editorial cartoon appeared (the source lost from memory) which depicted a long line of workers outside an unemployment office. Half way down the line stood a robot. An officer from the unemployment office was motioning the robot forward, shouting "Next!" The cartoon succinctly captures the insecurity many people feel about the potential effect of automation on their jobs. These feelings are not helped by misconceptions that are often fanned into flame by the popular press. Stories about the factories of the future—factories that need no lights, no heating, no windows, and no cafeteria because there are no human workers—are fascinating reading but do not reflect incremental conversions of existing manufacturing facilities.

Studies of employment levels are also difficult to conduct. We understand that technology, economic growth, and employment are all inextricably bound together, but a precise understanding of the relationships is difficult. Many factors come to bear on the problem of predicting employment effects: interest rates, inflation, economic growth. The increasing importance of foreign competition within the formula has added even greater complexity. American jobs may be exported overseas to take advantage of low-cost labor, but at the same time the standard of living in foreign countries may rise, creating a larger market for sales. Relationships between these factors are so complex that economists who study the trends and make predictions often rely on simplified assumptions which, if improperly employed, can distort the overall picture.

Hunt and Hunt (1983) cite one example of how this distortion can occur. They reviewed manufacturing employment projections from 1979 to 1990 calculated by others, based on two major assumptions, which concluded that 22 percent of all existing jobs in manufacturing could disappear by 1990. The assumptions were that productivity gains would remain at a constant rate of 2.1 percent annually and that output would remain constant. The latter assumption particularly is unrealistic, but oversimplifications of this type are typical in making such predictions.

What do we know about the impact of technology on employment levels? The question can best be answered by segregating the long-term and more immediate impacts.

Long-Term Effects on Employment Levels

Over the long haul, technology has a clear and undeniable effect on standard of living and employment. This issue can be put into perspective by examining the effects of the Industrial Revolution, which provides an historical role model for current developments. Work during the early days of industrialization was characterized by twelve-hour days, six-day weeks, and child labor. Retirement and medical benefits were nonexistent. Through productivity increases brought about by "hard" automation, the population gradually redistributed itself into different patterns of employment. A higher standard of living, shorter work schedules, elimination of child labor, medical programs, and retirement benefits were also part of the outcome. Today, the U.S. economy is near full employment, taking into account normal transitory unemployment and economic cycles.

The argument to be made is not that technology created unemployment among the workforce, but that it was the engine of prosperity that generated high levels of employment and a superior status of living for Americans unique in the world. According to Harvey Brooks:

> Economic historians have long pointed out that since the early nineteenth century, superior technology has been a major factor in America's competitive advantage in world trade and in its higher average real wage levels. Since 1880, U.S. manufacturing exports have been largely concentrated in new products that other countries either could not yet produce or had not yet begun to produce in quantity. To maintain its higher wage levels and living standards, the United States constantly has to develop new, unique products (or new types of tradable services) to exploit the temporary monopoly arising from their uniqueness: charging higher prices while not being faced with price competition. It is primarily this economic rent arising from innovation that enables U.S. producers to pay higher wages. By the time the new products or their production processes become standardized, and thus producible competitively in lower-wage countries, the United States would have developed improved versions of the products, more advanced production

> technologies, or entirely new products whose sales would be sufficient to offset the gradual loss of market for the more standardized items. In this theory, the rate of innovation in the United States must be fast enough to maintain and increase high-wage jobs more rapidly than they are lost throughout the transfer of older technologies abroad. (Scott and Lodge, 1985, p. 328)

What this means is that the American workforce is presented with two basic scenarios. First, if the rate of technological adaption and innovation is slow, the United States will continue to lose competitive advantage as it already has to nations such as Japan. American workers will take the brunt of this scenario in two ways. American jobs will be lost overseas to countries with relatively low wages. Or American workers can take substantial cuts to bring their standard of living into parity with citizens in competing nations. Neither option seems attractive when one considers that the world population is a seemingly bottomless pit of humanity, most of whom would be delighted to work at small fractions of U.S. wages.

The second scenario is that U.S. businesses restore their aggressive and innovative stance where the application of technology is concerned. This avenue offers the best solution for retaining a leading role in international competition, for keeping employment levels and the standard of living high. The downside is that some *temporary* displacement of the workforce is inevitable, an issue we shall return to shortly.

Management has two responsibilities with respect to the long-term employment issue. First, managers have a responsibility to influence public policy makers regarding the context in which businesses must be managed: the environmental context. Issues related to the environmental context are taken up in Chapter 9. The second responsibility relates to the organization's internal workforce and could best be characterized as an education or communications issue. As part of the ongoing effort to manage technology, management should constantly inform and remind employees of the role technology and innovation plays in competition. Despite the importance of technology in our economic history, the lesson has not been taken. According to Hunt and Hunt:

> There appears to be a significant lack of understanding that one of the consequences of a growing dynamic economy, one that makes more goods and services available to all of us through the productivity gains of its workers, is job displacement or the elimination

> of some jobs through technological change. Concomitantly, we know that other jobs are being created, sometimes in the very same firms that adopt new technologies and sometimes in altogether new sectors of the economy. (1983, p. ix).

Thus, if the economy is to grow and prosper, U.S. managers must be reconciled to technological change and the short-term impacts it will have on employment levels. Within this framework, we now turn to the question of the nature of these short-term effects.

Short-Term Effects on Employment Levels

Undoubtedly, adaption of new technology will create short-term structural unemployment. Structural unemployment occurs because of changes in the employment needs of society; some jobs become obsolete while new job categories come into demand. So far, most evidence suggests that the American workforce does not face massive unemployment for two reasons. First, the introduction of technology has been accompanied by economic growth so that job losses in one area are offset by gains in other areas. These offsets may or may not be in the same company, industry, or geographic region. The second reason is that technological implementation has gone much more slowly than most predictions. Technology experts and most managers seem to be eternal optimists when it comes to forecasting the speed of technological change.

Predictions about the impact of office automation have varied greatly. Yet penetration of technology into managerial and professional work categories continues to be relatively slow. In manufacturing, robotics and computer-integrated manufacturing are in their infancy and it is difficult to predict precisely what the employment effects will be. However, some research indicates that robot adoption will not have a significant impact on total manufacturing employment by 1990. In fact, Hunt and Hunt estimate that robotics will eliminate less than one percent of all jobs in this period.

Of course, the impact of automation is selective. Robots have penetrated automobile manufacturing more than any other U.S. industry. Production painters, the job category with the greatest impact, will experience a drop of one-fourth to one-third in the number of jobs by 1990. In other industries, only 7 to 12 percent will be displaced, bringing the overall average to 9 to 15 percent. Thus, certain job categories in certain industries bear the brunt of automation more than others.

We see the same selective impact in the services sector, with the heaviest declines in office jobs occurring in the financial services companies, particularly insurance. However, the employment impact of white-collar automation has been very difficult to pinpoint. As we saw in Chapter 5, the economics of these technologies, especially communications-based technologies, are only poorly understood. This problem is compounded by the fact that we understand little about the nature of white-collar work, especially at the managerial and professional levels.

Clerical work is the easiest to predict; we have already seen reductions in some industries (e.g., insurance) and foresee more to come. It is at the mid-level in the organization where changes are less well understood. Most observers agree that a reduction in traditional middle management is already underway. This group, which grew in force following World War II, was responsible for gathering and analyzing large quantities of data. On average, American firms came to have a disproportionately large number of middle managers when compared to major trading partners. Technology and the desire to slim organizational ranks to improve competitiveness will clearly have an impact on this group. However, the growth in the service sector and in knowledge work does much to offset any losses. By 1995, according to various predictions, the service sector will account for three-fourths of the U.S. economy. The greatest category of growth is technical and professional (i.e., knowledge work).

We come away from this discussion with two major conclusions. The first is that from a general economic perspective, even short-term displacement of workers is likely to be slight, and this may be negated by overall economic growth. This observation is based on our current understanding of technology's impacts. As we move further along with computer-integrated manufacturing and the automation of knowledge work, our conclusions may change and then a reassessment would be in order. But for the present the issue appears to revolve more around the question of what structural changes in employment will occur, rather than displacement of workers in general. The second conclusion is that the impact of job displacement when it does occur is selective; some industries, and some job categories within those industries, will feel it more than others. Just how much job displacement will be experienced is a question that should be addressed as part of the technology assessment procedure. Evidence suggests that displacement effects, if they are forecast early, can be offset through normal attrition.

The implications for management policy making fall into two categories: situations where there are displacement effects that can be contained internally within the company and where some layoffs must eventually occur.

Contained Displacement

Contained displacement results when the introduction of technology causes jobs to be lost in one area but the loss can be absorbed through attrition or transfer to a different job. IBM is a prime example of a company that practices this approach. Long noted for their bias against layoffs, IBM transferred approximately 25 percent of its total workforce from its Lexington, Kentucky plant when computer-integrated manufacturing was introduced; a reduction from 6,000 to 4,500. Although costs of transfer are high, because of moving expenses and retraining, most companies who practice this philosophy believe the costs work out to be less in the long run when hiring and termination costs are taken into account. Workers who remained at Lexington and took up new positions also had to be retrained, but that was considered a normal part of the cost required to convert the plant.

Whether an employee is being transferred to another company facility or retrained for a new job at the existing facility, education is the critical factor. It is the lubricant, so to speak, that makes all the parts work together in technology-based change. Companies also report that relocation and retraining of the workforce should be considered as early in the planning for new technology as feasible. As already noted, education and training issues should be routinely monitored as an ongoing part of the technology assessment effort.

Organized labor is, of course, a mitigating factor in plans to relocate and retrain the workforce, and generalizations about possible scenarios are impossible. Some unions, especially in industries where displacement is substantial or potentially so, have shown a willingness to work with corporate management. In other situations, union leaders have not shown much inclination to cooperate.

Layoffs

Circumstances sometimes force companies to lay off workers. However, management should be clear about what the circumstances are, because technology is often blamed for job displacement when poor performance of the economy (or industry or company) is really at fault. Hunt and Hunt cite one example:

> The U.S. economy recovered very slowly from the deep 1958–59 recession and then experienced another recession in 1961. The "automation problem" was of urgent national concern, and in 1962 the U.S. Congress passed the Manpower Development and Training Act to address the retraining needs of technologically displaced workers. Then, in 1964, the President appointed a National Commission on Technology, Automation, and Economic Progress to determine the impact of automation and technological change on the U.S. economy.
>
> But the economy was already beginning to recover significantly in 1964, and by the time the Commission rendered its final report in 1966, the economy was near full employment. Historical events ultimately obviated the need for and impact of the Commission; the problem seemed to have gone away. To no one's surprise, the Commission's conclusion was that a sluggish economy was the major cause of unemployment rather than automation. (1983, pp. 20–21)

Blaming technology, while it may seem to be an ideal scapegoat in the short term, has untoward consequences over the long haul; future implementations will have to live down the reputation technology has earned for displacing workers. Indeed, even in cases where technology is a defensive strategy against the incursion of foreign competitors, efforts should be made to educate the workforce to the tradeoffs that must be made in terms of jobs lost to technology or overseas.

Most American companies have a relatively unenlightened attitude towards technology-based job displacement. The major exceptions to this gross generalization are consumer-oriented companies or companies that are headquartered in small communities. Here the potential ill will created in the minds of consumers or the local community creates a sense of caution about layoffs and how to deal with them. Americans already change careers throughout their productive years more often than is commonly thought, and the rate of change is likely to increase now that certain job categories have a shorter life-cycle. Only a decade or so ago, computer programming was considered to be a promising career path, but today many programmers worry that they will be "buried in the basement" with the big computers. Thus, society will have to learn to deal with rapid job turnover in careers to an increasing extent. But part of the responsibility belongs to corporations who must lay off workers.

Assuming that job displacement cannot be contained within the company through transfer and attrition, companies who wish to maintain high standards as an employer have little choice but to provide potential layoffs with the opportunity to retrain for jobs in demand and outplacement services. These benefits still may not totally satisfy displaced workers who must leave family roots and hometown for geographic regions that offer better job opportunities, or who must face change at a time in life when they would prefer not to do so. To ameliorate these sometimes unavoidable consequences, a company might provide counseling services directed at helping the employees make the adjustment.

Again, when workers are organized, management's freedom to structure programs to ease and facilitate job displacement are proscribed. Numerous union leaders, when convinced that job displacement is inevitable, cooperate to see that the interests of their membership are served as best as possible under the circumstances. And some union leaders who foresee substantial displacement have proposed or implemented funding for job retraining when layoffs occur.

Restructuring the Workforce

In the final analysis, job displacement must be addressed at the public policy level for the simple reason that it belongs to a larger class of issues involving the restructuring of the entire workforce. Given the growth in service sector jobs and the shorter employment lifecycle, the wokforce is moving away from domination by committed, full-time employees to domination by "contingent" employees. The norm used to be that a worker expected to make a life commitment to a particular company and expected the company to return the commitment. Evidence suggests that we have begun to depart from this norm and that a growing number of employees are seeking work on a contingent basis. This work is generally of short duration and is often part-time. In 1986, 25 percent of the U.S. workforce (including willing part-timers) belonged to this contingent workforce. Contingent workers include individuals who work at home, who are involuntary part-timers, or who are employees of subcontractors, "leased" employees, or temporary workers.

The value of a contingent workforce is that employees can be added and subtracted when needed, thus increasing the flexibility of the economy and the company's ability to respond to change. The mix of work skills can be changed over time as needs change. The problems of obsoleted job skills and having a deal with resistance in the workforce can be largely avoided. If a contingent worker wishes to remain employable, he or she must be willing to adapt to change and keep skills current. As the skills the contingent worker brings to the job erode, or as the demand for these skills diminishes, the worker must enter a period of education and training to be competitive again. Despite the publicity the Japanese have received for lifelong employment practices, their workforce is largely contingent. In fact, permanent and contingent workers will often wear separate types of badges to identify their status.

Although many people in the United States prefer contingent status because it suits their lifestyle for one reason or another, being a contingent worker has its drawbacks. Contingent workers, especially part-timers, earn substantially less than their permanent counterparts. Seventy percent have no retirement benefits and 42 percent have no health benefits, nor generally are there plans they could elect if they wished. In Japan, cultural differences avoid many of these problems. Women, if they work, are almost always contingent workers. But because of the emphasis on the family in Japanese society, even if a woman is not employed, an income is still being derived from the male's employment in most cases. American society is, of course, on quite a different course aimed at equal employment opportunities for women who may or may not be married. Unfortunately, the historical biases against women often still have the effect of placing them in the contingent workforce against their will.

If contingent work is to be an enduring feature on the employment landscape, and one that facilitates technological change, responsiveness to changing market conditions, and revitalization of work skills, then support systems will have to be designed and put into place. Support systems would include retirement plans that work from employer to employer. For example, many universities and colleges use a common retirement system for educators: the Teachers Insurance and Annuity Association (TIAA). Although the rules for vestment vary somewhat, educators can transport their retirement benefits as they move from one institution to another, something they tend to do frequently. In the past many retirement systems were designed to encourage employee loyalty but now new systems are needed to facilitate contingent workers. Putting these systems into place will most likely require intervention from forces outside the corporate environment. For example, the TIAA was set up because of initiatives taken by the Mellon Foundation, which was interested in improving support for educators. Thus, the issue of how to deal with contingent workforce issues may belong to the public policy forum. Technology-related public policy issues are considered in Chapter 9.

The Transformation of Work

Information technologies are not only changing the structure of the workforce, but are also changing the nature of work itself at the individual and group levels. While job displacement addresses the question of what work is disappearing, transformation of work issues deal with where work is going. The management issues revolve around concerns about what jobs will exist in the future. Specifically, management must know what the requirements of these jobs will be, and who will succeed and who will fail in these new roles. In short, management must know what policies are needed to achieve a match between work skills under technology and available human resources. Bringing about major change in the mix of job qualifications is as difficult and time consuming as any other form or organizational change. Therefore, policies should be directed at the changes that will be required over the long term, perhaps five to ten years into the future. In that sense, transformation of work policies is closely tied with the issue of building general competitive capacity in the technology infrastructure. In the following pages, three major impacts of information technologies on work are considered: (1) educational preparation needed for work, (2) evaluation of work performance, and (3) methods of work performance.

Educational Preparation

The rapid rate of technological change, coupled with pressures from international competition to be more responsive to market demands and more flexible operationally, is creating a rising demand for education as a lifelong process. Lifelong education facilitates change in two ways. First, it prepares workers for new job opportunities when existing skills are no longer in demand; in other words, it increases worker mobility. Mobility may take the form of outplacement in the case of layoffs or of reassigning workers within the company as the need arises. However, for the foreseeable future technology has increased the likelihood that an employee will have many different assignments throughout his or her career. Second, it refurbishes knowledge needed by an employee preparing for the career phase ahead even if that means just bringing current work skills up to date. We have already seen in Chapter 7 that education is an important factor in breaking down organizational resistance to technology. Thus, the issue is not only training people to do new jobs or to do jobs differently with technology, but facilitating organizational change.

Building general competitive capacity with technology means making the appropriate organizational changes, and one of those changes is to translate human resource requirements into an educational policy viewed over the long term. Education is endemic to technological innovation within organizations. Unfortunately, today's companies are largely on their own in terms of solving the educational issue since traditional educational institutions—colleges and universities—have not been able to respond adequately to contemporary needs. Although American higher education has become much more involved in creating new knowledge through research since World War II and has sallied into the area of "adult" education through continuing education and executive programs, the system remains firmly rooted in the undergraduate liberal arts and professional schools approach. This approach says, in effect, that the principal role of education is to prepare the graduate for the beginning of his or her career. While there are some programs that respond to needs that arise later in one's career, they are the exception—presented outside the normal educational framework—rather than the rule.

Corporations now spend unprecedented sums on internal education, more than the traditional higher education system. Some even offer accredited college degree programs to their employees. They are attempting to address many of the needs that the traditional higher education system has not met. Managers moving into senior jobs face completely new problems and concerns, technology management being one example. Learning general management skills may be important to a professional who is making the transition from a narrow specialty to middle management. Boards of directors often need exposure to issues such as directorship liabilities and defensive strategies in take-overs. Yet, much of the structure of current business education is targeted to the undergraduate and MBA, basically addressed to the entry-level job market. The public policy concerns associated with needs for lifelong education are taken up more fully in Chapter 9.

Some readers may feel that a discussion of lifelong education issues is remote from the central question of managing information technologies. But the intent of this discussion is to point out that because education is often relegated to a position of secondary importance by managers, it never fulfills its potential as a facilitating function in organizational change. And without appropriate organizational change, the technology is quite likely to perform below expectations. Within the context of the Integrated Technology Management Framework, education of workers—the process of preparing them for new roles—is equally as important as the investment decision, understanding technology, or any other management responsibility.

Evaluation of Work Performance

Much of our current philosophy of managing people is based on physical observation of work activities. In fact, the organizational concept of "span of control"—the number of people a single individual can effectively oversee—is a direct product of this approach, as is the practice of keeping workers in close proximity to their supervisor when arranging workspaces. Increasingly, however, growing use of telecommunications is permitting employees to work away from the traditional office environment most or part of their time, a trend referred to as "telecommuting." Some observers have forecast a return to a cottage industry system in which people work at home and stay in contact via telecommunications links.

Telecommuting is likely to offer some flexibility in allowing workers to redesign their work schedules in a way that benefits both employee and employer. When charged with preparing a report, an employee might do so away from the office to avoid interruption. Some workers are, of course, housebound due to handicaps, child rearing, and other reasons. Many could be productive wage earners if they could find employment that did not require their physical presence at a traditional worksite. And, we have already noted that telecommunications permits companies to move "back-office" operations away from costly city centers and will undoubtedly, with time, allow them to be moved around the world to take advantage of inexpensive service labor. For many workers, telecommuting offers lifestyle advantages; it avoids the necessity of wearing business attire and permits relative personal freedom. Other people have difficulty in working without constant direction and may feel socially isolated. Yet, despite drawbacks, telecommuting is likely to become more common.

Methods of Work Performance

Information technologies are also having an impact on how work is performed at all organizational levels. As we have noted throughout this book, these impacts are not revolutionary, but evolutionary, slowly molding work habits to be in sync with the opportunities offered by the technology. Almost since the beginning, observers have been speculating on the impact computers would have on management. We have already observed, for example, that during the 1970s technologists believed that most of the information management needed to operate the business could be provided through totally integrated management information systems, a belief that proved to be fallacious. Many have also debated the virtues of real-time access to data bases and whether or not managers could or could not use terminals to answer most of their questions about the business. Yet we have also seen in our discussion on the economics of technology in Chapter 5 that management is a highly variable activity. Different people use different approaches at different times for different tasks, suggesting that technology's impact is likely to be uneven at best. This appears to be especially true at higher levels in the organization. The truth is that at senior levels, direct use of the computer is likely to remain mostly a question of personal choice.

Providing Leadership

At the middle management level, automation's impact is likely to be more widespread and managers are likely to have much less discretionary choice about using it. Part of the reason, as we have seen, is that the information collection and analysis functions handled at this level lend themselves to automation. Another reason is that office automation generates certain productivity gains only when a critical mass of users is reached. Most of these gains flow from communications-based technologies: electronic mail, calendaring, voice messaging, electronic filing. Thus, to permit managers to remain outside the boundaries of automation is, in effect, to turn one's back on potential benefits.

The implication of this is that all managers must be on the communications network. Not being on the network is like not having a telephone. No matter how relevant or irrelevant a manager considers the computer to be to his or her job, no matter how alien to his or her personal syle, the manager must still be on the network. If the boss is not on, subordinates are not likely to be on either. But, if the boss makes it clear that he or she intends to communicate through his medium in the future, then subordinates are likely to be much more receptive. One top technology executive used this bit of psychology to install a voice-mail system in his division of about three thousand people. His strategy was to begin using the system himself, telling his subordinates that if they wanted to communicate with him, they must use voice-mail. Soon they joined, followed by their underlings, and so forth until virtually the entire division had been converted to the new system, reaching the critical mass necessary to derive the benefits. Had a different strategy been used, one beginning at the grassroots level, resistance to the new technology and familiarity with old ways of communicating would have persisted, as experiences in other divisions of the same company proved.

The voice-mail case provides a compelling example of how policy about use of computer technology in management jobs can have a major impact on the acceptance of new technologies, especially at the senior management level. But, two caveats need to be added. First, we have had some experience in the past with management providing a "role model" to workers by having a computer on the desk, an approach that met with mixed success. What is different about the voice-mail example is that it enforced the use of the technology, it did not simply set up a role model to be emulated. The second caveat concerns the sincerity of the effort. Managers who use technology in their jobs as mere window dressing are quickly discovered by subordinates and the leadership value of the gesture is understandably diluted. Does this mean that senior managers must also be hands-on with the technology? The answer is "no." A secretary could as easily handle the dealings with the machine as long as the message is clear: the communication system—not paper—is the network.

Two Schools of Management

A chorus of voices is growing within many companies, insisting that personal choice among managers about use of information technologies is not realistic. Technology has, they argue, so fundamentally changed how management should be done that those who are not using technology are simply not managing. Thus, technology has in a real sense created a dichotomy of thinking in management. On one hand, there is the taditional school of thought which views technology as having a limited and specialized role within organizations. On the other hand, technology-oriented managers understand that the fundamental precepts of business management have changed.

According to the latter school, managers must understand how to use technology to "manage by the numbers," as an executive in one company described it. Management-by-the-numbers logic is simple. The contemporary business is so large and complex that data rather than direct observation is the primary basis for decision making. The problem is that companies now produce far more information than management can reasonably digest. Managers often complain about getting too much data and having trouble judging what is relevant to the business and what is not. The higher one goes in the organization, the more difficult the problem is likely to be. According to John F. Rockart:

> General managers—and especially chief executive officers—whose needs for information are not as clearly determined as are those of many functional managers and first-line supervisors. Once one gets above the functional level, there is a wide variety of information that one might possibly need, and each functional specialty has an interest in "feeding" particular data to a general manager. (1979, p. 81)

The problem has been that we have long assumed that the technology would be able to process data into meaningful information for consumption by management; this is embodied in the cliché that we do not need more data, but better ways of processing data. It is true that technologies such as expert systems will assist in the task, but the technology school argues that technology cannot do it all. Managers themselves must know what numbers are important or, as one executive phrased it, "they must know how to find the pulse of the business." Managers must then know how to use the technology to retrieve and manipulate the relevant data. Whether a manager personally uses a computer to obtain the data is irrelevant. The point is that he or she must understand the workings of the technology in order to be able to accomplish the job at hand. Many managers will use the technology directly. One executive told of a new division manager: "The first day he arrived on the job, he demanded that numbers important to him on a daily basis be put on his personal computer."

The policy issue confronting managers is how they will support the new technology-oriented school of thought. Without support, through words or deeds, senior managers effectively suppress productivity gains by discouraging or even ignoring new management techniques that rely on computer technology. Young men and women moving up the ladder take their cues from their superiors, and the message should be that the ability to exploit new technology is looked upon favorably even if that was not a factor in senior managers' careers. Evidence suggests that creating a climate where using technology is encouraged produces many unexpected benefits. According to David Davis:

> The most interesting effect is that work on [personal computers] is so cheap and quick that managers are willing to undertake speculative or exploratory exercises. Since it is difficult or impossible to judge the outcome of such work in advance, the work was simply not done on mainframes as it required an investment of scarce programming time. (1984, p. 65)

In essence, what experience demonstrates to us is that many benefits from information technology are unforeseen. They are realized only when energy is invested in actually using the technology as an element of management. Without technology, the technology school argues, managers simply lack the appropriate tools to get at the relevant data. And until they are actually using the technology, they cannot gain experience or develop personal capacity in this arena nor can they hope to enjoy the unexpected benefits.

9

The Environmental Context

Outside Factors

Managers must face one final reality when managing information technologies: that the ability to use technology effectively depends on factors outside the organization and outside their immediate sphere of influence. Dealing with external constraints is as much a part of management's responsibility with respect to technology as working within the internal constraints. Thus, it is not enough to understand and assess new and emerging technologies, nor to divine the economic behavior and plan the technological infrastructure, nor make investment decisions, nor create a climate for organizational change and prepare for the transformation of the workforce. All these factors are necessary, of course, as previous chapters have described. And they are all generally accepted as the types of activities that managers should and would act to control. But they are not enough.

In recent years, I have often observed management teams spending significant time and effort to redesign management practices and control systems to improve the results they are experiencing with computer technology. While these efforts may produce some benefits, many are doomed from the start because they do not resolve the most important problem, a problem that might be termed the "environmental context." By this I mean constraints that are placed on the ability of American firms to compete. More specifically, for purposes of this book, we are interested in outside factors that adversely affect technology-related decisions. Most of the major problems with technology management would be resolved if these external constraints did not exist or were different. But given the current state of affairs, managers are left without any external support for making technology-related decisions that would benefit the company in the long term. And managers who make appropriate technology investments too often do so at great risk, putting their credibility on the line and exposing the company to possible take-over action.

One responsibility of senior management is to be active outside the company in creating a healthy climate for business—in this case as it relates to managing technology. Thus, the Integrated Technology Management Framework includes the management of factors in the external environment as a component equal in importance to any other. For purposes of discussion in this chapter, these factors are grouped into three categories: (1) the accounting and financial community, (2) industrial policy, and (3) business education and research.

The Accounting and Financial Community

In earlier chapters, we explored the biases inherent in American accounting standards and how they have been derived from the self-preservationist instincts of the profession since the 1930s. We have examined how these biases distort perceptions of the financial health of the business over the long run by emphasizing short-term performance measures. We traced the roots of these short-term measures to find that they have come to create relentless pressure on managers to produce results in the immediate quarters at the expense of long-term investments in general competitive capacity. We have judged that this system has constrained managers from making decisions based on the economic realities of technology rather than the distorted perspective of financial accounting standards.

Internal accounting and financial controls have also failed management. Cost accounting methods used by most firms today are at least two to three decades old and are based on economic assumptions that are no longer relevant. According to Johnson and Kaplan:

> The obsolescence of most companies' cost accounting and management control systems is particularly unfortunate for the global competition of the 1980s. The consequences of inaccurate product costs and poor accounting systems for process control and performance measurement were not severe during the 1970s since a combination of high inflation and a weak dollar sheltered most U.S. manufacturers from foreign competition. High levels of worldwide demand for U.S. products during that decade placed a premium on production throughput. Higher costs and, occasionally, goods of substandard quality could generally be passed on to customers.
>
> The competitive environment for U.S. manufacturers completely changed during the first half of the 1980s. First, disinflation reversed the previous inflationary psychology, and manufacturers could no longer recover cost increases through higher prices. At the same time, a sharp increase in the value of the U.S. dollar made foreign produced goods less expensive to U.S. consumers and made U.S. produced goods more expensive to foreign purchasers. In addition to these dramatic changes in the macroeconomic environment, a revolution in the organization and the technology of manufacturing operations was leaving laggard U.S. producers in an even more precarious position.
>
> At first, U.S. manufacturers believed that the inroads being made in their traditional markets by foreign, particularly Japanese, manufacturers could be attributed to lower overseas wage costs. Only with some delay did they recognize the onset of a revolution in manufacturing operations. The revolution was triggered by innovative practices developed by Japanese manufacturers during the 1970s and by the availability of new technology that greatly reduced the direct labor content of manufactured goods. Leading the new revolution were new practices emphasizing total quality control, just-in-time inventory systems, and computer-integrated-manufacturing systems. (1987, pp. 209–210)

In summary, accounting measures provide the wrong signals to managers when it comes to technology-related decisions. At a time when American companies need to be making long-term investments in technology infrastructure in both office and factory, accounting emphasizes the short term. At a time when the fundamental economic and cost behavior of technology-based production has shifted away from direct labor, cost accounting systems have become all but irrelevant. And at a time when manufacturing and office automation has become for many companies the single largest component of the capital budget, accounting offers no new insights about how to measure long-term capacity.

Observers frequently argue that all this makes little difference since the financial community is sophisticated enough to recognize when management makes investments that brighten the long-term prospects of the company. Aside from the fact that there is little evidence to substantiate this argument, it really misses the point. The point is why the American system of accounting controls emphasizes short-term performance criteria if the financial community wishes to encourage long-term success.

The best evidence available suggests that building technology infrastructure takes time and must be done incrementally to be successful. The conversion of the IBM Lexington plant, for example, took approximately six years and could still not be regarded as totally complete. Building the new General Motors' Saturn plant will take even longer before the flow of benefits begins. Current measures simply cannot cope with a time horizon of this magnitude, at least not in terms of giving formal recognition in the financial statements of the enterprise.

How then can the situation be changed to support effective decision making with respect to technology? The answer lies in several options open to management.

Designing New Internal Measures

The most readily available option available to management is to begin redesigning internal management control systems that take new economic realities of technology's performance into account. For example, traditional systems have relied on mass inspection of incoming materials and outgoing products to determine whether the system's performance is within "acceptable" limits. But it turns out that emphasis on quality—achieving zero defects—encourages constant attention to improving manufacturing processes, eliminates the need for inspection and rework, and generally causes total manufacturing costs to decline. Such results can lead to very different conclusions about how management should invest in technology infrastructure.

Reforming Accounting Measurement

Managers also have a clear responsibility to work with the accounting profession to help reshape both the system of financial accounting reporting practices and methods for internal performance measurement. Judging the extent of the reform needed is well beyond the scope of this book, but the task will undoubtedly be substantial. One can glean some idea of the scope by considering the extent of the process needed for changing current standards, including dealing with the vested interests of the accounting and financial community as well as the educational effort that would follow. Even electronic data bases, systems of measuring performance for credit purposes, and numerous other activities are based on current measures. Change in accounting and financial standards, like any other change, would have to be gradual to be successful.

Reexamining the Importance of Income

In the final analysis, redesigning internal reporting systems and reforming accounting measurement standards may still fall short of the mark when one considers the financial and economic structure of Japan and other international competitors. Japan's financial system almost completely by-passes short-term income measures, focusing instead on market share, growth, and total sales. As already noted in Chapter 5, this system allows management to reinvest in future capacity. Gains on investments are in the form of long-term capital appreciation. Indeed, relatively high levels of corporate income tax further bias managements' decisions in favor of investments in technology. The important point is that the financial structure and method of performance evaluation are consistent with economic realities. Long-term investments in technology infrastructure are encouraged.

Many readers will, of course, have misgivings about whether challenging the traditional income concept is relevant to managing information technologies. Those readers may wish to consider that one of the effects of having stronger foreign competitors is that they tend to invest more heavily in United States corporations. Most experts generally agree that foreign investments increase interdependence between nations and provide jobs for American workers. However, when income from those investments is taken as dividends (i.e., is not reinvested in the business) by foreign owners who plow all such proceeds into future capacity, the stakes change somewhat. Not only is the fruit of American production not reinvested for the future, but it goes to another country to be invested for *their* future.

Many react by proposing protectionist measures. Yet protectionism, while it has some limited uses, stifles trade in the long term to everyone's loss. Protectionist sentiments do not address the underlying problems that make our accounting and financial systems indifferent or even hostile to our own welfare. Many other examples exist. Foreign banks operating in the United States, for example, are allowed to combine many services such as merchant and investment banking, which American banks have been forbidden to do so. The point is that the inadequacy of accounting measures may simply be symptomatic of the deficiencies in the underlying financial structure.

Industrial Policy

The relationship between American business and government has always been one of laissez faire, with little governmental interference in economic affairs except for minimal maintenance of peace and property rights. The American military has been known to step in to protect U.S. corporate interests abroad, and the government has played various regulatory roles: the Securities Acts were enacted to protect citizens in transactions related to public investments. It has also acted in various ways to facilitate American business at home and abroad, and has guaranteed corporate finances when the public interest was at stake.

But the U.S. government has never played an active leadership role in directing industrial strategy, with the possible exception of war-related efforts. Nor, given the economic dominance of U.S. business throughout the world, has it ever had to play such a role. Indeed, the American culture is deeply rooted in beliefs about the freedom of action. The U.S. business community generally prefers to be left to its own devices, remaining deeply distrustful of governmental involvement. It comes as little surprise, then, that the United States has little in the way of national industrial policy. According to Albert Sommers:

> The U.S. has virtually nothing in the way of a coherent industrial policy. Unlike most of our international competitors, we tend to leave the market adjustment process alone, even if it is moving at a perilously slow rate, and even if it exposes us to gross inefficiencies and painful unemployment. (1983, p. 22)

We cannot ignore the fact, however, that other governments play leadership roles in their nations' business affairs. There is no doubt, for example that a strong industrial policy has been of critical importance to Japan's rapid economic rise. Japan's policy attempts to anticipate major developments, including those which are technology-related, and it encourages businesses to involve themselves as soon as possible. Two examples will suffice. One is the push by Nippon Telegraph and Telephone, with government support, to create the Information Network System, a nationwide digital telecommunications system. The second is the Fifth Generation Project, Japan's push to create true machine intelligence. The Fifth Generation Project was launched as a national initiative comparable to the United States mission to send a man to the moon. It is, in the words of one Japanese researcher, an effort to "redesign the computer."

Japan's government has also provided leadership for advances in new technology through dramatic efforts to build a series of high-tech cities throughout the country, the so-called technopolis concept. Although the practice of building new cities is deeply rooted in Japanese history, the effort still must be appreciated for its commitment to very long-term economic development. In Japan, these policies are attempts to catalyze economic growth by supplementing and complementing market forces. In other words, the industrial policy does everything within reason to facilitate the commercialization of technology.

A major vehicle for implementing this policy in Japan is the Ministry of International Trade and Industry (MITI). This organization was reformed in 1948 and is responsible for making industrial policy and managing foreign trade and commercial relations. In addition it runs the patent office and facilitates the acquisition of raw materials, energy, and other resources. Through MITI and a network of related organizations, Japan provides soft guidance to its industries by circulating information and ideas and helping shape corporate strategy. The primary goal is to provide orderly markets focused on *competing with businesses in other nations* as opposed to competing internally with each other.

This shift of focus to foreign competition may force the U.S. business community to reevaluate the desirability of having some type of coordination of American economic efforts. Where information technologies are concerned, the primary value of national policies would be to facilitate market processes, encouraging greater clarity of strategic purpose. Sommers (1983) describes sound industrial policy as having purposes:

(a) [that attempt] to accelerate healthy adjustments in markets that would otherwise be achieved only over a painfully prolonged period;

(b) that seek to achieve outcomes that markets cannot hope to achieve under existing laws (for example, the antitrust laws);
(c) that underwrite, by guarantee or by actual carrying of the costs, outcomes that markets themselves will not produce under any circumstances, and
(d) that seek to counteract the effects of industrial policies pursued by some of our trading partners, without resorting to overt protectionism.

Whatever form a national industrial policy takes in this country, it will have to be appropriate to the American culture, resolving the need for coordinated efforts yet avoiding unwanted social planning aspects. Such a solution is not likely to be found by emulating Japan's or any other country's system. According to John A. Dierdorff, Managing Editor of *Business Week,* "We are missing the point in concentrating our attention on competition in Japan." Managers, he notes, are as susceptible as anyone to "leaps of fashion." Here, one recalls with some irony the recent flow of American managers to Japan on observation missions when only a few years before, Japanese businessmen were typically the ones journeying to the United States. What Dierdorff meant, of course, is that the solution does not necessarily lie in copying other production systems but in American managers providing creative solutions to real problems. This logic extends to national policy as well. The objective is not to recreate industrial policies of other nations on American soil, but to create a policy framework that would be conducive to technological innovation.

Frank Press, President of the National Academy of Sciences, has created a simple national innovation matrix (Figure 9-1) for "assess-

	Science & Technology Policy	Macroeconomic Policy	Management Skills	Sociotechnical Factors
Research				
Development				
Production				
Distribution				

Figure 9-1. The National Innovation Matrix—A Framework for Comparing National Policies and Capacities for Technological Innovation.

From A High Technology Gap? Europe, America, and Japan *by Frank Press, p. 17. Reprinted by permission of the Council on Foreign Relations, copyright 1987*

ing national policies and culture as they might influence innovative capacity." The rows of the matrix represent the classic innovation cycle: research, development, production, and distribution. The columns refer to important national policies and attitudes: science and technology policy, macroeconomic policy, management skills, and sociotechnical factors. Technological innovation can occur at any point within the matrix. As Press notes:

> As the Japanese have shown in many cases of commercial success, a solution to a difficult development problem rather than a research breakthrough, or improvements in the manufacturing process to increase productivity or quality, can enhance the competitiveness of a product. Often, an improved product is more successful than an innovated one. Salesmen may be the first to perceive the need for a new product, which then triggers new research and development efforts. (1987, p. 17)

Numerous proposals have been put forth over the years to deal with technological innovation. Many simply repeat previous proposals so that now the list has become more or less standard. Using Press's framework, following are discussions of some of the major proposals.

Science and Technology Policy

The purpose of science and technology policy is to focus the objectives of technological innovation in research, development, production, and distribution in ways that enhance competitive capacity. The following proposals fall into this category:

- Direct commercialization of new technology
- Creation of a federal science and technology department
- Removal of regulations and antitrust barriers
- Increased support for research and development
- Protection of intellectual and property rights, and a streamlined patent process
- Revised trade laws and policies

DIRECT COMMERCIALIZATION OF NEW TECHNOLOGY

Even if the United States government does not take a direct hand with private industry in developing commercial strategies, it does so indirectly. In the 1960s, for example, when President John F.

Kennedy challenged the nation to send a man to the moon by the end of the decade, much of his motive centered on the need to increase spending in a laconic economy. Numerous scientific and technological developments generated for the space program ultimately found their way into the commercial sector. The Reagan administration's Strategic Defense Initiative provides a current example. The primary objective of "Star Wars" is military and political, to create a space-based high technology defense shield that is particularly effective at the negotiating table with the Soviets.

However, given proposals to spend billions each year for research and development, spinoffs into civilian applications are likely to be large in number. These spinoffs include energy, production, transportation, medicine, and communications, among many areas. Indeed, this symbiosis between Star Wars and commercial technology extends to most defense spending. Universities with research and development grants have the opportunity to refurbish and upgrade their facilities. Private companies take technologies developed for defense and convert them to commercial applications. Large consulting companies have developed specific organizational mechanisms for extending experience gained in defense-related work to the private sector.

This process of spawning commercial applications of technologies developed for defense is pervasive. A regional telephone company, for example, developed a market research system based on electronically mapped data. Age, income, marital status, and other factors were coded by geographic region within certain cities. Electronic mapping allowed age to be combined with income, for example, to produce a visual map of the market profile. However, the company also combined the electronically mapped market data with an expert system designed to perform the analyses. The development question for the company was how to build such an expert system: Who had the appropriate experience? The answer turned out to be the Defense Department, which was developing battlefield management systems based on artificial intelligence technology. Battlefield management turns out to be basically a question of interpreting electronically mapped data, the same conceptual issue facing the marketers at the telephone company.

But a debate exists about whether expenditures on defense-related goals with indirect commercial benefits are enough. The conversion of technological innovations from defense to private sector is inefficient. Writing for the *New York Times,* Malcolm Browne summarized the argument as follows:

> Critics of S.D.I. point out that the technological side benefits of Star Wars research could be had much more cheaply and efficiently if they were pursued directly rather than as the unintended offshoots of an extravagant military spending program. But S.D.I. proponents assert that in the absence of such a visionary scheme, it is unlikely that such research would have taken place at all. (August 24, 1986)

The issue is not whether technologies developed originally for defense purposes should be commercialized or not. Rather, the issue is whether a separate program should exist which has the promotion of technological innovation for direct commercial interests as its primary goal. The exact commercial priorities would have to be established only after examining where the United States' competitive strengths and weaknesses lie in the emerging international scheme of things. Again, American managers should keep firmly in mind the growing competitive prowess of nations such as Japan, the fact that they do not currently have significant military expenditures as does the United States, and that they pour significant resources into developing technological initiatives such as machine intelligence, supercomputing, and digital telecommunications networks.

Creation of a Federal Science and Technology Department

One of the proposals advanced by the 1985 report of the President's Commission on Industrial Competitiveness is that a cabinet-level department of science and technology be created "to promote national interest in and policies for research and technological innovation." As envisioned, the department would serve as an overall coordinating agency to improve the effectiveness of research and development in meeting long-term goals. Most importantly, where information technology is concerned, it would provide management with a medium to express their concerns and gain the benefit of others' thinking. The department would be the keeper of United States national policies and would serve much the same role that the MITI does in Japan. In administering these tasks, the organization should, in Sommers' words:

> . . . aim at the reinforcement of investment trends where there is a clear concensus of need; it should take an administrative form resistant to simple appeals or help from any one quarter. In some

> degree, it would displace the violent and often uneconomic competition among state and local governments to attract investment. It would represent, at best a national effort to supplement market forces in stimulating and shaping the future capital stock. (1983, p. 23)

A department of science and technology could also play a pivotal role in reversing remnants of government laws and policies targeted to antitrust, swinging the balance away from regulation and control to encouraging competitiveness.

Companies have used various mechanisms to try to overcome the lack of coordinated efforts in American industry. For example, the Microcomputer Corporation (MCC) located in Austin, Texas was established by a consortium of firms to do research on problems related to microelectronic technology. Other companies have tried to bolster the effectiveness of their technological research efforts through joint ventures. While these enterprises are often successful, many are plagued by problems that reduce their overall effectiveness. Companies are reluctant to share technical information for competitive reasons. The venture may turn out to be a drain on cash, producing no immediate profit. One example is a joint venture between several leading banks and a telecommunications company which was established to prepare for electronic home banking, a market that is expected to mature slowly and therefore not generate immediate profits. Yet, the CEO of the joint venture reported that within a year or two of setting up the venture, partners began to press for business activities that would produce immediate income.

Finally, a science and technology department could also play a useful role in setting standards for computer technologies, especially in the communications area. As we have already seen, many benefits of office automation and computer-integrated manufacturing depend directly on all technology components being able to communicate. Current problems with incompatibility pose significant barriers to the efficient solution of these problems. The barriers are less technical in nature; more political and commercial. Properly administered, the ability to facilitate standard setting and implementation on a broad scale would justify establishing an agency.

Removal of regulatory and antitrust barriers

Many current laws are vestiges of earlier concerns in our history for unfair competitive practices; others relate to the regulatory era that followed the 1930s Depression. Often these laws and regulations do more to hinder American business than help it. We have already noted, for example, that foreign banks have more freedom in U.S. markets than do American banks. Especially where disincentives to engage in joint research are in evidence, regulations need revision.

PROTECTION OF INTELLECTUAL AND PROPERTY RIGHTS

Greater regulation and enforcement is needed to protect patents, copyrights, and other intellectual property rights from counterfeiting both in the United States and abroad. Patent laws should also be revised to further encourage technological research and development, for example by extending the patent life lost during the government's approval process.

REVISED TRADE LAWS AND POLICIES

Recent developments in international competition have created a need to review domestic trade laws and policies regarding international trade. Especially important is the issue of dealing with unfair trade practices of foreign governments and the criteria under which revised laws should protect U.S. industries from outside competition. Closely associated is the need to revise the global monetary system.

Macroeconomic Policy

The second set of policies identified by the Press matrix relates to macroeconomic issues. In general, one can argue that any policy which supports business ipso facto is good for technological innovation and should be encouraged by managers wishing to improve their performance of information technologies. For example, the 1985 report of the President's Commission on Industrial Competitiveness recommends a reduction of the budget deficit, a stable money supply, and tax reforms so that "the efficiency with which resources are allocated is improved and the cost of capital is lowered."

Specifically, two proposals are worth noting. The first concerns needed reforms in the financial community regarding the focus on short-term performance. This issue was, of course, discussed earlier in the chapter. The 1985 President's Commission went about as far as any other proposal by recommending that:

> Money managers, bankers, accountants, stockholders, and business leaders should be challenged to deemphasize simple short-term financial measures more consistent with long-range competitiveness and profitability. (1985, p. 52)

But the question of providing the appropriate framework for managerial decision making with respect to technological innovation is so critical that a recommendation of much greater strength is indicated.

The second proposal concerns providing economic incentives to increase research and development activities both in universities and private business. These incentives would take the form of favorable tax treatment and increased government funding. Providing additional support for research and development is important because many projects are too risky to undertake as purely private ventures. Thus, some mechanism is needed to offset or share this risk to insure that firms have a reasonable chance of being able to earn a return on their investments.

Management Skills

One area that has been almost completely ignored or that has received only cursory attention has been management. Frank Press suggested in his national innovation matrix that management skills when coupled with an educated and dedicated workforce can work together constructively and can adapt easily to new technology. But, the issue is much deeper. In the first chapter, I argued that Japan's economic successes can be traced in large part to management's willingness to adapt their thinking to challenges at hand rather than the country's ability to carry out technological innovation per se. Indeed, Japanese management's adaptability was related less to technological innovation than to the ability to respond to a real problem with a creative solution. Early efforts to compete globally were based directly on areas where they were likely to be most successful:

> Starting with "smokestack" industry products and following with high-technology products, Japan has emphasized up to now the later stages of the innovation cycle, i.e., the lower half of the national innovation matrix [see Figure 9–1]. Highest priority was given to putting in place the most advanced process and production technology, aggressive pricing, and mass marketing. Japan was content to be the second to the market with new products believing—rightly so—that better quality and lower costs of its products would displace those of the original innovators. It confirmed with a vengeance the comment that "pioneers are those with arrows in their backs." (Press, 1987, p. 23)

What can be said of management as an institution and the way managers relate to technology? First, according to Peter G. W. Keen, "Managers don't know much about computers" (1985, p. 35)—an observation confirmed by executives themselves who responded to the Fordham survey. Second, we know that the most lucrative career options often reside in finance and law rather than in production, marketing, and administration. This bias in the mechanism that effectively allocates the nation's best talent further contributes to a U.S. business system that is ineffective in formulating creative solutions to real problems. Third, executive compensation is often tied to accounting and other measures which, as we have already seen, contain serious distortions. A fourth and final point is that management lacks an effective support system that promotes excellence and quality among its own ranks, an issue I will return to shortly. This lack is acutely obvious in the area of information technologies management. Many managers continue to concern themselves with basic technical questions, not realizing that the real problems are not technical at all. According to Peter G. W. Keen:

> Although technocal innovation and design are still important, information technology has moved out of that domain and now is part of the structure of everyday life and of organizational and social innovation and design. It has moved into the domain of values: What sort of an organization is desired, and what will the result be if particular choices about [information technologies] are made? Yet no established tradition exists for helping manager set policy for [information technologies]. (1985, p. 38)

Sociotechnical Factors

The fourth area of concern for national policy identified in Press's matrix is sociotechnical factors. These factors encompass a broad list of concerns including level of R&D sophistication, responses of labor and bureaucracy, and the cultural inclination to save, assume risk, and provide social support.

Business Education and Research

The 1985 report of the President's Commission on Industrial Competitiveness noted that universities are under stress in many areas which could contribute to the nation's competitive health. Specifically, they noted that:

> University revenues do not cover the rising cost of research, and engineering faculty salaries do not compete with those of private industries. As a result, fully one-tenth of the nation's engineering

> faculty positions are currently vacant. In critical fields like electrical engineering and computer science, some universities report that half of their positions are unfilled. As the American Electronics Association stated so well, we are "eating our seed corn" by failing to produce the faculty to produce future scientists and engineers. Our scanty harvest shows the results; Japan produces more engineers than we do. (1985, p. 20)

The commission recommended that the educational process be improved in four ways. First, in engineering education, support should be encouraged for adequate faculty salaries and additional funding should be made available to continue engineering research and to upgrade as well as maintain equipment and instrumentation. Second, the commission suggested a partnership between the federal government and the private sector to counter the high dropout rate in schools. Third, technology should be used as a means in itself to improve the educational process. And fourth, the commission recommended that business schools be "challenged to undertake a systematic and comprehensive academic response to the changing competitive environment." It is this last recommendation that is relevant here.

Many business executives have come to feel that American business education is seriously deficient—part of the problem rather than the solution. Sheldon Zalaznick, managing editor of *Forbes Magazine,* was asked about the value of academic business research to his publication. He responded with three quotations. The first was from Hermann Goering, the Nazi Reichsmarschall, who said "When I hear the word art, I want to reach for my revolver." Likewise, said Zalaznick, "When I see the scholarship bearing on productivity techniques, I want to reach for a revolver." The second quote is from Woody Allen's movie *Love and Death.* Allen, who plays a court jester who tries to seduce the queen, is fumbling with her chastity belt, fearing that the king will momentarily return. "We must hurry," he says, "the Renaissance is coming and then we'll all be painting" meaning of course, that unless some action's taken soon, the opportunities will be lost. Finally, Zalaznick quoted Ralph Cortiner, former chairman of General Electric. Cortiner was responsible for carrying out a plan to decentralize General Electric. In an interview Zalaznick asked him why he had done it. Cortiner replied: "If you take over a company that is centralized, then it's probably a good idea to decentralize it, and if you take over a company that's decentralized, its probably a good idea to centralize it." The point,

Zalaznick concluded, is that good management is the product of intelligence, common sense, and, hopefully, decent intentions. The rest is commentary and most commentary, he continued, is boring and unhelpful. Much of what passes for helpful insight isn't.

From an educational perspective, business schools have a somewhat uneven reputation. MBA programs are now pumping out more than 70,000 graduates per year. And despite the complaints voiced in private by many managers that graduates "can't do anything," demand has generally remained strong and has increased with time. At the high end of the scale, graduates of elite schools and even top graduates in second-tier schools receive lucrative offers. Competition for these graduates can be intense. Whatever its shortcomings, many firms still believe that the educational system has some value as a mechanism for funneling desirable job candidates their way.

At this point, we must ask ourselves two critical questions. The first is whether business educational institutions can provide assistance in the effort to better manage technology? The second is whether business schools have a role to play? I believe they do for three reasons: (1) they provide quality education, (2) they provide valuable research facilities, and (3) they serve as demonstration centers for new technologies.

Quality of Education

Despite complaints regarding the quality of business education and business school graduates, companies still hire graduates in record numbers. The debate about the value of an MBA has raged for years. Yet for the best students job offers at entry level can be lucrative. And, the total number of MBA graduates has grown steadily since the introduction of the degree. Whatever the shortcomings, companies have come to feel that some exposure to the complexities of the business environment is better than none at all. Thus, these companies are willing to pay at least some premium for people with a business education. In a sense, what businesses face is a sunk cost. Given the fact that the premium is paid, there would seem to be an incentive for increasing at least the relevancy and quality of these programs.

Think Tanks

As we said earlier, some companies such as The Travelers are beginning to experiment in research and development programs that pertain specifically to computer-based technologies, following the tradition of the industrial research budget. Yet much of the experimentation could be done at universities. Like the centers for excellence or technology centers discussed in Chapter 4, universities should be serving a role as technology assessors and transfer agents. Just as companies need to become more aggressive, universities need to find mechanisms to carve out a role for themselves in this area.

The relevance of this to business education and technology management is that the technologies being created in this environment are those that business students will manage in the future. Although there is no traditional role for such development in business curriculums, the time has come to link this type of creative effort with business education. The reason is simple: the ability to manage the commercialization of a technology is intimately tied to its development. This includes understanding the nature of breakthrough and "hitting walls." Up until now, these issues have been regarded as the domain of engineering majors, not business students. But managers now oversee technology and must understand the related processes. The failure to move these issues into business schools perpetuates the mentality that technology is something best left to technical people. Thus, business schools have an important role to play in moving the technology process closer to those who will manage it in the future.

Demonstration Centers

Finally, business schools have an important role to play as demonstration centers for new technologies. Far too many educational institutions, almost all, concern themselves almost solely with the problem of providing exposure to the most basic computing capabilities. While this "bread and butter" issue must, of course, be addressed first, additional emphasis on other, more advanced and experimental technologies must be included. Without these extras, students develop only a myopic vision of what information technologies are about. A more complete idea of technology capabilities is needed.

This myopia is present in the business faculties as well. Too many schools have taken the view that managing information technologies is really a question of performing quantitative techniques on the computer. Not surprisingly, these schools usually have strong quantitative and operations research oriented faculties already in place.

The infusion of a new perspective into business programs is critical. Most schools find themselves grappling with questions of basic microcomputer facilities for students forever caught up in a cycle of obsolesence and replacement. Their energies are never mustered for the main event, which is to introduce a much richer perspective of the technology. As a result most business schools trail behind developments in the private sector. The opposite should be true; they should be providing leadership in the technology area.

Many recommendations have been forthcoming about what should be done to improve American business education. These recommendations are usually ambiguous, agreeing on little more than that improvement is needed. Most critics agree that faculty salaries should be increased and additional funds made available for research and new equipment. But something more is necessary: aggressive reform of business education. Increased resources are undoubtedly needed, but the structure that consumes those resources must also be made more effective. Most importantly, the reform of business education must include mechanisms to ensure that business schools take a leadership role in information technology management. Their role in improving national competitiveness would be a key one. Properly organized, business schools should become the crossroads of new and developing technologies, a place to exchange information about corporate problems and needs, and a center for research capable of providing truly paradigmatic thinking. In other words, business schools should serve many of the functions performed by coordinating agencies in other countries.

10

Principles for Managing Technology

Regaining the Cup

When I set out to write this book, I did so for one major reason: managers were being given so much disparate advice about how to manage information technologies that it all sounded like garble. Often the advice was contradictory. Often it counseled radical change based on new methods of management without explaining how or why the change should be made. These revolutionary management methods were predicated on notions about human organization that ignored the most obvious realities: for example, that hierarchy and bureaucracy are necessary for communication and control. And often—far too often—the promises made about productivity gains and strategic advantage simply could not be validated by experience.

There was a need for a conceptual framework, I felt, which would provide senior managers with a starting place or unified perspective in thinking about information technologies. The framework would allow managers to clarify their role in managing technology. But more importantly, the framework would provide managers with the ability to better evaluate various advice and counsel they received both from internal and external sources. Is the advice, for example, consistent with the realities of what we know about the role of hierarchy in human organization? Does the timing of benefits promised from office automation take into account the need for major social change within the organization's culture? Is the effort to assess new and emerging technologies consistent with the importance of information technologies to the overall competitiveness of the firm? And so forth.

Originally, the idea for building the Integrated Technology Management Framework was to gather together the various commonly accepted ideas, concepts, and practices and then attempt to piece them together into a whole. That effort was only partially satisfactory and it quickly became obvious that large pieces of the puzzle were missing and that more questions were being asked than answered by the exercise. For example, in light of all the emphasis on using technology for strategic purposes and advice that managers must become more "involved" with technology, why are there no specific mechanisms (i.e., technology assessment) to deal with these issues? Why do experts keep promising significant improvements in white-collar productivity when office automation has historically not produced the expected benefits? And how can managers make wise decisions about investments in technology systems when the tools for evaluation are lacking and the outside financial influences flagrantly contradict the realities of technology's economic performance? And why do experts continue to focus on the presumed negative impact of technology when experience shows that people often have positive, even powerful, feeling towards technology?

Because our thinking has been dominated by ideas that do not reflect economic realities, two dysfunctional consequences have occurred. First, many American managers have come to believe that they cannot trust their experience and instincts when it comes to technology. They cannot, for example, build complex computer-integrated manufacturing systems to reduce costs in the long haul because they would be reducing profits in the short run. They are told by experts that they need new ways to manage, but, no one seems able to articulate adequately what the new ways are. Secondly, this fractionated, unreliable perspective produces a situation where American managers seem to be fumbling around looking to make sense out of the computer "revolution."

These situations combine to make the American manager appear both uninformed and incompetent and that, I believe, is the root of the anxiety many of the nation's senior-most executives feel when confronted with technology issues. Many would argue that managers are now past computer anxiety. Yet, in my various travels, I have had occasion to talk to many of the lions of American industry, and almost to the man, their roar becomes deft side-stepping when the issue of information technologies arises. The anxiety is alive and well.

On the positive side of things, we have learned much about how to manage information technologies over the past thirty years, and especially in the last decade. Indeed, if it were not for this rapidly accumulating body of knowledge and experience, the task of developing a conceptual framework would be impossible. The time is ripe to inventory that knowledge or—given its lack of organization—to integrate it into a unified perspective. This, too, is the basis for the Integrated Technology Management Framework proposed in the book.

To start, the Framework had to speak to issues of valid concern to senior managers, and that mandated that it not be technical in nature. It also had to be general enough so that managers could adapt it to a broad range of industries and firms. And it had to be flexible enough to accommodate rapidly evolving technology. Three major premises formed the foundation of the Framework. The first is that fundamental changes in the international marketplace are causing the bases of competition to shift and American companies are having to learn to compete in unfamiliar ways. Competition driven by foreign forces is increasingly based on satisfying customers by offering quality in products and services. This means that quality must be built into production and administration practices from the beginning as a means of producing quality output. But, quality also lowers costs. Productivity gains come from reducing rework and defective products. Information technologies are a critical resource for competing successfully in this new environment. Yet this resource has been undermanaged and improperly managed, with the result that businesses have not adopted computer technologies as rapidly as possible, and when they have been adopted, the outcome has often been disappointing.

One might ask why American managers have not been more proactive in managing technology considering the growing competitive threat from abroad? The answer offered in this book is that the historical evolution of accounting performance measures and other factors removed important incentives to engage in technological innovation. Many of the support systems were also allowed to deteriorate; examples include the research infrastructure and the education system. And circumstances did not challenge managers to respond to the strategic opportunities offered by computer technologies. Since the early 1980s, recognition of these problems has been growing and some evidence suggests that despite limitation, American managers are beginning to change course. However, many observers question whether American industry can recover its footing, or whether the seeds of inevitable decline have already been sown. Only time will tell, of course, but it might be worth noting the striking analogy the 1987 America's Cup Race bears to our current competitive situation.

After twenty-four races spanning 132 years in which the United States consistently won the America's Cup Race, Australia defeated the Americans in 1984. Technology was responsible for the Aussie win. Their boat, *Australia II,* was designed with a revulutionary winged keel that helped solve two problems in sailing: it provided lift to reduce the drag on the hull while still keeping the weight low to maintain the vessel's balance. Where the Americans had seemingly resigned themselves to the limitations in twelve-meter design, the Australians had used technology to achieve a quantum leap in performance. In doing so they essentially changed the basis of competition in the America's Cup Race.

One cannot help thinking about the speed with which nations such as Japan an South Korea brought about a quantum leap in their own economic performance as well as a new basis of competition in the international marketplace. In the same way that the Australians dealt a stunning defeat to American supremacy in yachting, these nations have dealt some startling blows to certain U.S. industries such as automobiles and electronics. As already noted, the United States has lost leadership in seven out of ten high-technology industries and the distance is narrowing in others. Just as the 1984 America's Cup Race forced us to admit that someone else had designed a better twelve-meter yacht, so must we admit that in international competition, others have demonstrated a superior strategy for the times. In both contests, technology has played a central role.

The American yacht that won the 1987 America's Cup Race—*Stars and Stripes*—was also the product of technology. At least one participant described the competition as being one of design and technology, not sailing. Following the 1984 defeat, a design team was formed under American yachtsman John Marshall which included a number of technology specialists. Making exhaustive use of computer-assisted design (CAD), graphics, and other applications including computer-assisted tank testing, Sail America (the organization that built *Stars and Stripes*) built three different yachts. After the first two designs performed only adequately, the design team came to believe that in order to win, another boat was needed, one that would represent in itself a major advance in design and performance. In eight tight months, working under tremendous pressure and the knowledge that the Australians were not resting on their laurels, the team designed, built, and tested the new boat. The rest is history, of course.

It would be incorrect to credit technology alone with the victory. But using technology in an environment of risk taking and innovation would come much closer to the formula that accounted for the success. I don't believe that the formula should be any different for American businesses currently facing international competition.

There is a downside, a potential pitfall, to this parable. One might be led to the conclusion that when a competitor in yachting, commercial enterprise, or whatever sneaks up from behind and steals a temporary advantage, all that is needed to recapture victory is the good ol' college try. Did we not come back to win the war over both the Germans and Japanese following a surprise attack at Pearl Harbor? Did we not trounce the Australians in their own waters off Perth to regain America's Cup? Indeed, American flexibility and ingenuity is revered by many not only in the United States but overseas as well.

There are, however, limits to where flexibility and ingenuity can take us. Nowhere is this more obvious than in information technologies, where we have already lost the lead in many high-tech areas. The absence of a coherent industrial policy in the United States that would help set goals for commercial technology development, and the absence of effective policies and support for research contrast starkly with the Japanese effort to build the next generation of computers. Since the early 1980s, Japanese government and industry has been fervently working on a carefully planned group of projects which will start being combined in the waning years of this decade. Individually these projects are impressive enough; when brought together the synergy will be significant. Development of this new generation, based on "knowledge information processing" is a national priority. It is viewed in Japan as the key to future economic survival and success. It is where the Japanese have chosen to make their mark.

Making predictions about the course of information technologies is a practice fraught with risk and little reward, but I will indulge anyway. Mark your calendars. By the mid 1990s, barring major unforeseen events, Japan will lead the world in virtually every aspect of information technologies. To remain competitive, much less dominate, the United States would have not only to revise its stance with respect to industrial policy but reform basic financial conventions. At this juncture, there is no organized support for doing either; indeed there is little awareness or agreement that these changes are necessary. Yet from a strictly technology-oriented point of view, if nothing else, these changes should be priority policy concerns for the U.S. president who takes office in 1989.

These issues fit into a larger class of concerns about how competitive the United States can be in international markets. This book has left little doubt about my position on this subject. It is disconcerting, though, to hear that many managers have become bored with these issues and tired of hearing about them. Jaded politicians in Washington, one leading newspaper reported, refer to the issue of competitiveness as the "C-word." Such attitudes do not help; they do not solve the substantial problems before us.

The second major premise of the Integrated Technology Management Framework is that we have reached a level of understanding and maturity in our dealings with computer-based technologies which should give us answers to questions about how to manage them. Or at least we should be in a better position to ask the right questions, such as why isn't the process of adopting technology faster and more successful? And why haven't many of the expected benefits materialized? The bits and pieces of the puzzle are largely available. We understand the major technical trends in information technology's evolution, even if we do have a tendency to exaggerate their futuristic effects. We have a methodology for assessing new and emerging technologies and infusing the resulting knowledge into the organization. We have evidence about the economic effects of information technologies though our understanding is imperfect and needs refining. We now see that technology is intimately tied to the competitive performance of the company and has the potential to change the basis of competition within industries and internationally, an approach that increasingly focuses on satisfying the customer through quality. Emphasis on cost cutting per se may accomplish little. We know that to be successful, technological change must be driven by appropriate changes in the organizational culture. These cultural changes must in turn be driven by the competitive goals of the business. We know that the workforce must be prepared for technological change and that the bonds between humans and machines can be strong. And, finally we know that the signals managers receive from the external environment create strong disincentives for making effective technology-related decisions.

What remains is to put the pieces together in a coherent fashion that helps managers articulate their responsibilities. Hopefully, the Integrated Technology Management Framework presented in this book has made a contribution in this area. The Framework is meant to deal with technology management issues at the executive level. As such, it should help senior managers determine what is important and what is not. The thrust is to understand technology (assessment), decide its uses (position-taking), and set the stage for implementation (policy formulation) by subordinates. But the Framework is also designed to be a frame of reference for *technical* management. Issues like information architecture planning, project management, systems analysis and design, computer auditing, and computer security that have occupied a place of central importance in management thought about computing in the past now take their rightful place as secondary management concerns. Secondary means that they come into play only after senior management establishes the appropriate policy framework. The existence of a policy framework that deals with broad business concerns rather than specific technical issues is critical as information technologies become more decentralized and more under the control of nontechnical business unit managers and other end users.

The third premise is that information technology management should begin with a traditional orientation to management's roles and responsibilities. Or, perhaps more to the point, there must be an understanding of traditional roles and responsibilities. Managers provide leadership and values to the company, and set direction. Managers decide resource allocations. Managers motivate employees. Managers are emissaries to outside organizations that influence the company's activities. Managers balance innovation and control.

This same orientation and understanding can be extended to the organization itself. The fundamental nature of human interactions that form the basis for human organization is not likely to evaporate because we now have global telecommunications networks and desktop computers. Hierarchy and bureaucracy will continue to govern the organizational form. The capacity for communicating more freely and directly, which technology gives us, does not alter basic human motives for limiting and directing the flow of communications. Such motives are often valid and the results necessary to maintain organization.

Certainly, changes will continue to occur as they have occurred in the past. Companies will become more responsive and flexible. Organization structure will flatten for a time. Communications will become more fluid. Emphasis will be placed on innovation and creativity at a time when innovation and creativity are needed. Computer technology is indeed a powerful vehicle for facilitating these changes. *But,* let us never overlook the fundamental question about whether such changes are needed and desirable in specific circumstances because in many cases they are not. Because technology can take us there does not mean we want to go.

The Principles

The Integrated Technology Management Framework provides a broad outline of the major issues and the progression of those issues when managing information technologies. But it does not necessarily provide directives or principles for transforming management's perspective. What does the Framework say to management? How must managers do their jobs differently if they expect to improve the results obtained from information technologies? To answer these questions, the following ten points distill the important points or lessons from our discussion and attempt to pinpoint the key issues in managing technology.

1. *Accept management's role in technological innovation.*

The process of managing technology begins at the top. Observers have long exhorted managers to be involved or play sponsorship roles in technological innovation and implementation. That is not enough. Direction in information technologies must emanate from the senior level which defines business strategy, allocates investment funds, and establishes organizational policy. There are two fundamental reasons for this. First, in many cases computer technologies have become such an integral part of the business that they cannot be segregated from the business objectives. Technology leads to new products and services, and existing products and services can be given new life through technological applications. As one Fordham survey respondent said, "People think technology is something separate from the business; it is the business!" For some companies that rely less on technology, this manager overstates the case. But as already noted, when a firm does not use technology as a competitive tool, it should be the result of a conscious decision and not by default due to management's ignorance or mismanagement.

The second reason management must accept its technology-related responsibilities is because by not doing so these tasks default to others within the company, often people in charge of technical management of computer resources. The responsibility does not go away; decision making has simply been unwittingly been left to subordinates who are not generally capable of understanding corporate strategy issues and therefore cannot make the appropriate decisions. Indeed, they do not know they are making strategic decisions that affect far more than the technical issues they believe themselves to be addressing.

According to Peter G. W. Keen:

> Because delegation has for so long been used in implementing [information technologies], top managers are now sanctioning radical organizational change without recognizing that they are doing so. Office technology is just one example of investments being made in the name of productivity with no clear picture of how the organization might, would, or should be different as a result. There is a vast amount of literature on the unintended consequences—not always negative ones—of office technology and on the naive views of work, people, and productivity that too often lead to the transformation of technical successes into organizational failures. (1985, p. 36)

Ultimately, the largest concern facing the United States today is how to get senior managers into the pilot's seat where technological innovation and implementation are concerned. Managers have generally been sluggish in responding to the need for technological change. Or, more accurately, they have been slow to recognize that the basis of competition is being altered by foreign companies.

The tragedy for American enterprise, and one of the important lessons we have gleaned from our experience with information technologies, is that it takes time to build technology infrastructure. The complexities of new technology systems, especially those based on large-scale integration as in computer-integrated manufacturing and office automation, cannot be constructed and generate benefits overnight. And because evidence suggests that some benefits will not flow from the systems until a critical mass is reached, many companies may be in for a long wait for changes that will make them genuinely competitive, even if they began in earnest developing these systems today.

American management must therefore sweep its own doorstep first. As an initial step, resistance among senior executives should be dealt with in a forthright way. Excuses that "managers resist because they lack keyboard skills," that age is a factor in microcomputer acceptance, and similar caveats should be dispatched for the palaver they are. What is nonnegotiable is that managers must change their own attitudes and philosophical outlook on their roles and responsibilities. If educational and other support systems are not in place, create them. If traditional university business schools cannot respond to the need for quality education and research, and the need to be creative and innovative with technology, then abandon them and establish a private system that will.

2. *Commit equal energies to each and every phase of technology management: assessment, position-taking, and policy formulation. None ranks above the other in importance; none is technical.*

The Integrated Technology Management Framework outlines the major issues that demand senior management involvement. What can we conclude from the concerns that are included and those that are excluded? First, technical issues are almost completely absent. Although the assessment phase is largely concerned with management's understanding of information technology, the emphasis is on the capabilities of new and emerging technologies and how they are likely to affect industry structure and competitive practices. Assessment does not imply becoming *technically competent* per se. Indeed, the more fundamental concern relates to the technology assessment methodology and information transfer mechanisms, and their appropriateness to the organization's needs. In position-taking, the same is true with respect to the economics of technology. As Chapter 5 noted, our lack of understanding about the economic consequences in white-collar work, for example, relates less to the technology than to the nature of the work itself. And, finally, policy formulation relates almost entirely to social and individual worker issues: how to accomplish organizational change brought about by technological innovation and implementation, for example.

Ordinarily, when an issue becomes technical, it signals a warning to senior managers that it should be delegated to the technical staff. Of course there are times when a senior manager must become involved in technical detail. For example, understanding a certain technical issue may be critical in making a decision about a trade-off in costs over the long term. Managers who oversee technical units—examples are telecommunications and central data processing—obviously have a greater need to delve into the nitty-gritty. But generally technicalities should put up a red flag that prompts a manager to question whether or not these details are appropriate to his or her interests. If there is any special concern with technical management within the company from an executive perspective, it is selecting competent, qualified people to supervise these areas.

For the issues included in the Integrated Technology Management Framework, each has the same importance as any other. This is, of course, untrue at any specific point in time since energies may be focused on one concern or another. But, it should be the departure point in thinking about technology management. In the book assessment was discussed first, position-taking second, and policy formulation third for presentation purposes. But in reality all three phases are ongoing activities that must be managed simultaneously. One could logically talk about organizational change (policy formulation) first, technological evolution (assessment) second, and investment decisions (position-taking) third.

Essentially, the Integrated Technology Management Framework subordinates technology to traditional business concerns. Technology assessment is, after all, no more than a specialized segment of the general monitoring of the outside business environment. Position-taking attempts to clarify investment decisions as they relate to information technologies. And, policy formulation considers the specific demands placed on commonplace organizational processes by technological innovation and implementation. Not only is the Integrated Technology Management Framework an effort to focus managers on important technology questions, but it attempts to do so in the context of traditional business issues. Thus, the *integration* in integrated technology management refers to the closure implied in managing computer technology as an indigenous element of the corporate system, not the process of combining technical components.

3. *Create a "vision" for the technology infrastructure based on general competitive capacity required for long-term competition.*

We have observed throughout the book that a fundamental shift has occurred in the economic behavior of information technologies. In earlier decades, integrating one computer application with another made little sense from a technical, economic, or business perspective. Tasks were automated piecemeal and each automation project was judged on its individual merits. The benefits that accrued from this approach were relatively easy to measure. But with the growing importance of *communications technology,* the emphasis shifted away from benefits that could be obtained by automating isolated applications and toward benefits that could only be obtained by *systems-wide* automation. In the factory, computer-integrated manufacturing systems produce the most lucrative economic gains by squeezing out excess inventory. Likewise, in the office, computer-integrated systems squeeze out unproductive time spent by white-collar workers, but only once *all* employees have been made a part of the system.

Perhaps because it is a physical process, managers have had an easier time in understanding the benefits of integrating manufacturing using a computer. Even so, we have seen how converting a plant from traditional production methods to a computer-integrated approach takes several years, five or more. Even with understanding, *a long-term commitment is required.* The benefits of office automation are more difficult to comprehend. The abstract quality of white-collar work combined with our poor understanding of knowledge work has created no clear recognition of how a traditional office system must be converted into one integrated by computers, following the example set by computer-integrated manufacturing.

In either case—factory or office—management must have a vision of how the technology infrastructure is going to look at a given future point in time. What it should look like, of course, is determined by whatever is needed to fulfill, technologically speaking, the general competitive requirements of the firm. Lead time is necessary to build general competitive capacity in the infrastructure. It is needed, as we have seen, because large and complex integrated systems must be constructed incrementally to be successful and even then undertaking their construction often entails considerable risk. Attempting giant leaps from one level of technological capacity to another as a strategy is not only risky, it courts disaster. One simply cannot count on technological breakthroughs occurring upon demand.

What this means is that investments in computer technologies can no longer be tied to explicit cost justifications of the type that discount *assumed* cash flows over the next five years or so. First, as we have just seen, the most important benefits may not materialize for several years. Second, one could attempt to deal with this longer time horizon by simply extending the cost justification into the future but most managers would agree that discounted cash flow is fanciful enough in the short-term. Underlying assumptions about interest rates or other economic conditions five or seven years in the future is pure speculation. And third, while management in many firms might be able to make reasonable predictions about trends in their industry five years or more into the future, precious few will be able to describe the specific competitive opportunities they will face.

Building technology infrastructure for general competitive capacity means just that. International Business Machines has built its Lexington plant (among others) to manufacture a broad range of products within a defined envelope; The Travelers Companies has developed a "data processing factory" to produce a diverse, rapidly changing mix of financial services products; McGraw-Hill has developed "platforms" for delivering "media blind" information. What is implied in each of these approaches is a sense of ambiguity by design: *general competitive capacity.*

Unfortunately, the need to build future technological capacity using an incremental approach that may not see the important benefits for some time places American managers on the horns of a dilemma. Every signal that they receive, whether it emanates from routine accounting reporting requirements or threats (real or imagined) from corporate raiders, reinforces the short-term perspective. Long-term initiatives that reduce current returns are taken *at risk.* Despite this risk, many American managers are still able to make the necessary commitments. But many are not. This is the most critical problem we face in managing technology because it supersedes the process of decision making in any particular American company. In many cases, managers cannot make the appropriate decisions in any situation.

4. *Build the technology infrastructure to produce quality products and services and build quality into the infrastructure. Continuously improve quality.*

The traditional view of information technology's role is to cut costs or improve productivity. Much of this view came directly from the limited perspective of the technical people who had charge of computers in the earlier days. They understood little of corporate strategy and therefore did not make a connection between the machine and the larger purposes of the organization. As a general rule, senior managers also failed in this respect. Perhaps it was because they regarded technology with the same limited perspective as the technologists, or perhaps it was because strategic thinking was not as important three decades ago as it is today. In any event, early cost reductions and other benefits were often lucrative and there seemed to be little to complain about.

In recent years, we have come to the recognition that computers have a broader role to play. Although productivity improvement is still important, many managers have come to see it as the handmaiden of corporate strategy. Improving productivity presumably allows the company to lower costs and that means being a low-cost producer, one approach to competition. Managers are also becoming aware that information technologies have other uses such as creating new or differentiated products and services. In fact, information technologies can, we are told, potentially affect any aspect of the value chain. Thus, our thinking has evolved to the point where we can say that information technologies improve productivity *and* differentiate products and services *and* allow the exploitation of specific market niches. But, that perspective is wrong because it still places the cart before the horse.

One must begin with the market niche. The technology system simply permits product and service differentiation as the customer demands. The route for satisfying that demand is quality. Quality not only distinguishes competitors on the field, it also provides the basis for lower cost through reduction of defects and reworking. The pivotal role played by the technology infrastructure is that a system can be built for quality or inferior production. But once built, the nature of the production is likely to be an indelible feature of that system forevermore. In other words, technology systems *institutionalize* quality or inferiority as the case may be. Thus, management can frequently make the largest impact on its competitive position by focusing its efforts on building a quality technology infrastructure. In a sense, the infrastructure clones that effort over and over in routine production of products and services.

A similar argument can be made for management's attitudes about the ongoing operation of the technology infrastructure. Even when major changes in strategic direction are not influencing technology related infrastructure, additional investments should be made to enhance the general quality or capacity of the system. The target is not achieving an "acceptable" level of performance but achieving zero defects and flawless performance: *continuous improvements.* Each time the system is improved, that level of increased performance or quality is replicated again and again.

Quality can be incorporated into human systems, of course. But as most managers know, unless cultural values constantly reinforce this perspective, atrophy can be rapid. Here, information technologies offer us an important added bonus, for as we observed in Chapter 8, technology systems can have a powerful impact on people's feelings. Technology can be a great communicator of corporate values and attitudes. It can, in a sense, "tutor" the human.

Whether one is considering the relationship between technology and the products and services it produces, or the relationship between technology and the humans who interact with it, the conclusion is the same. All are indistinguishable in terms of the orientation towards or against quality. To repeat an old Japanese poem quoted by W. Edwards Deming (1982, p. 177) that deserves repeating:

> *Kane ga naru ka ya*
> *Shumoku ga naru ka*
> *Kane to shumoku no ai ga naru*
>
> Is it the bell that rings,
> Is it the hammer that rings,
> Or is it the meeting of the two that rings?

5. *Manage technology as an investment, not a cost. Create a technology investment portfolio that balances the need to build general competitive capacity with short-term and medium-term exploitation.*

Simply building the future capacity of the technology infrastructure is not sufficient to realize the benefits being sought, of course. Over short- and medium-term time horizons, the general competitive capacity must be exploited to achieve specific business objectives. If the infrastructure has not been built to the appropriate capacity, needless to say that exploitation cannot occur. But just as too little emphasis on building capacity can shortchange competitive ability, too much emphasis on long-term capacity building can cause short-term opportunities to be by-passed. Thus, the objective with technology investments is to strike a balance between long-term and short-term needs.

Increasingly, short-term exploitation is coming under the control of business units rather than technologists and is expressing itself in two ways. First, business managers still retain the option of relying on technical support units to assist in developing applications where the skills needed exceed the capabilities of their own organizations. Typically, these situations are medium-range (three to five years). Normally, the application is developed through formal systems methodologies. But increasingly, business units are making direct use of the infrastructure's capacity through growing end-user competence without getting technical people involved. Because the capacity exists—in the form of data bases, accessible computing power, and communications—use can be made of it. This second approach is particularly effective for short-term needs. From an investment perspective, some allocation of resources is needed between building general capacity, an activity usually out of the control of most business units, and short- and medium-term exploitation.

Managers have another motive for diversifying their investments between long-term capacity and short-term exploitation. As we have already seen, capacity building may require a long-term deferral of major benefits which reflects unfavorably on short-term returns. Thus, by striking a balance managers ensure that some proportion of the returns on their investments find their way to the income statement.

6. *Use appropriate standards to measure technology's benefits, not just accepted standards. If you must, develop the standards yourself.*

Change in accounting and financial practices, if it comes at all, is likely to come slowly. In the meantime, distortions in performance measures will continue to influence management's decision making. Cost accounting systems were developed in an era when direct labor costs were a significant component of cost and indirect costs were considered relatively insignificant. As a result, cost allocation schemes were developed. But in many cases allocated costs now exceed direct labor costs, making the figures less meaningful. Allocations themselves contain distortions since they are based on historical costs which are in no way a reflection of value or a measure of productive capacity. As cost behavior shifts to become more fixed, less variable in nature and direct labor becomes an even less significant component, the distortion increases.

Financial accounting standards work in tandem with management accounting practices to reinforce and heighten the distortion. Allocation of indirect product costs is required practice. Through a number of mechanisms such as earning per share, periodic and segment reporting, and reliance on historical cost, tremendous pressure is brought to bear on American managers to pay attention to short-term profitability at the expense of long-term competitive viability. These mechanisms developed, as we have seen, because of a series of events in the financial history of the United States and the accounting profession's response to subsequent regulatory actions. Clearly, significant reform in the accounting and financial community is needed to develop measures that reflect contemporary economic realities. Specifically, some benefits from technology systems sometimes begin to flow only near the end of a long-term, incremental implementation process.

However, we may reach a juncture in our economic history where developing new and enlightened accounting measures is not sufficient to correct the problems. We now face foreign competitors who have effectively changed the rules of competition even for customers within our own country. Many of these foreign companies enjoy advantages that American business has never had and never needed: government support, low capital costs, and a financial structure that does not emphasize the need for income. The Japanese, as we have already noted, evaluate corporate performance based on total sales, growth, and market share. Cash inflows can be substantially reinvested in capital assets.

Is income an historical anomaly in the United States? Would it have decreased in importance as the separation of ownership and management widened in the first half of this century, had it not been for the interruption of the Depression? Did the 1929 Stock Market Crash and subsequent Depression create a regulatory backlash that in turn caused the accounting profession to react by institutionalizing income as a generally accepted principle? Did this institutionalization occur precisely at a time when income would otherwise have been waning in importance? And, finally, what is the value of income as a measure in judging the economic performance of corporations today? These issues go well beyond the scope of this book but they deeply affect the decisions discussed herein.

Aside from initiating widespread social and economic reforms, what can managers do in a more immediate and practical vein to alleviate the problem? The prognosis is not good. But one approach would be to aggressively develop supplementary measures that attempt to explain other dimensions of business performance. These might include long-term competitive capacity, current value of competitive capacity, and similar approaches aimed at telling a more complete story about performance. Managers face no prohibition in reporting these additional measures in corporate annual reports outside of audited portions unless it is to their competitive disadvantage.

Would such measures simply be window dressing or would they make a difference? Some evidence suggests the latter. Now that institutional investors account for such a large percentage of stock ownership in many companies, speculative trading in shares is difficult because of the sheer size of the investments involved. Some managers of these funds have come to believe that the major way to obtain long-term increases in their portfolios is through increases in the value of the companies themselves. They have thus begun to place increasing pressure on senior management to make decisions that are in the best interest of long-term shareholder value.

7. *Work to change the social, educational, political, and economic forces that shape managerial decision making with respect to information technologies.*

Constraints on effective technology-related decision making go beyond the accounting and financial community. As we saw in Chapter 9, technological innovation is affected by numerous factors in law and government policy. Many laws and regulatory practices are remnants of an era that concerned itself with antitrust activities. These activities were necessary to preserve the free market character in domestic competition. But the growing emphasis on international competition, or competition with foreign companies, has changed that need to one of providing leadership in developing competitive capacities. Technological innovation and implementation is chief among these concerns. Essentially, the nature of economic activity in the world is so changed by recent developments that we must rethink the roles and purposes of most of our social institutions.

Nowhere is this directive more important than in education, especially business education and research. Whatever their current shortcomings, these institutions have a real responsibility to provide a useful exchange of information on management issues between companies, governments, and other institutions. In addition, if properly administered they should serve leadership roles in creating excellence in management practices. I have no quarrel with people who claim that our present educational institutions do not serve that role. But the needs for superior thinking in management, the dissemination of information and knowledge, and education remain.

8. *Technology follows organization; organization follows mission.*

In Chapter 3 we observed that early centralized computers constrained normal organizational processes. Business priorities were subjugated to technical considerations. Long development and turnaround times for information systems were the norm. And large computers often had a centralizing influence on the form of the organization. But, with time this centralized environment began to erode as lower computing costs, telecommunications, and interactive processing began to distribute computing. Essentially, distributed processing has come to mean that an organization can configure its computing resources in almost any form or fashion needed. If a company is centralized, technology can be centralized. If a company is decentralized, technology can be decentralized.

The distribution of technology has also democratized it. The computing capabilities that could only be found in a large, expensive mainframe environment ten to fifteen years ago can now be found on a professional's desk. This has led many observers to conclude that because many computing tasks *can be* done at the individual level, that they *should be* done at this level. This has led to predictions about network organizations, work without bosses, and the electronic cottage industry. What these observers imply in their predictions is that technology will determine the form of the organization.

The form an organization takes is one response to the tasks it faces. If quick responses are needed, the organization is likely to be centralized for immediate mobilization. If the organization needs a way to contain the effects of mistakes and withstand political pressures, then a decentralized, bureaucratic structure may serve best. The form the technology infrastructure takes should follow suit and should match the organizational form.

9. *Establish appropriate mechanisms inside and outside the organization to assess new and emerging technologies and to transfer the resulting knowledge into the organization as a whole.*

Any period of rapid change, whether or not it concerns technology, increases the demand for more information. For years managers have been accused of knowing too little about information technologies, and by and large the criticism is valid. Undoubtedly, increasing the level of management knowledge and sophistication in this area would decrease resistance to technological innovation and implementation. In the past, the view was that this could be handled through education and training, and the infusion of younger, technology savvy managers. But the results of such efforts have been spotty at best.

The job of monitoring new and emerging technologies is not insignificant and cannot be handled in the casual fashion described above. For firms that rely heavily on technology, especially to enhance competitive advantage, a more systematic method of monitoring new technological developments and assessing their potential impact is needed. In other words, the current approach needs to be strengthened by formalizing and institutionalizing the gathering of information about technology and transferring that knowledge to all parts of the organization. The immediate impact of this action would be to reduce the cacophony of technical data that bombards the organization into useful, well-structured information.

As technology becomes more important to the firm, the stakes in understanding its potential uses increase sharply for two reasons. First, as we have already noted, technology systems often take time to implement. In the case of computer-integrated manufacturing and office automation, it may take years before systems are completely in place and benefits being obtained. To understand the technologies that will be available to that environment five or so years into the future, one must know what is being developed and tested in the laboratories today. The second reason is that firms are unlikely to be able to use technologies that are commonly available as a significant competitive weapon. Software or hardware available to one firm is available to all, making it difficult for a firm to sustain a competitive lead.

What this means is that many U.S. companies need to move farther upstream in their dealings with information technologies. Instead of being a passive consumer of hardware and software products, they need to become more involved with assessing research and development efforts by vendors, universities, and others. In some cases, this may lead to active involvement in developing aspects of the technology that promise to be of particular relevance to the firm at some future point in time.

10. *Use technology as a positive force in the transformation of work. Prepare the workforce for change.*

"Integrated" in Integrated Technology Management Framework also implies that technology and other parts of organization must be seen as a single system rather than separable components. Choices about technology are also choices about people. According to the Framework, failing to make decisions that relate to the workforce is just as prone to produce disappointing results from the technology as making the wrong decision.

The crux of the matter is that people have ambivalent feelings about technology. Most managers attempt to implement technology as if it were value free. But, as we observed in Chapter 8 it is not. Thus, technology will activate human reactions whether it is regarded by management as value free or not. As we also have seen, the problem goes much deeper than that because even when conscious choices are made about how information systems will interact with people, the effects cannot always be uniformly positive. Thus, intentional design of technology systems is important because it allows management to make trade-offs in terms of the human impact rather than producing unintended and unwanted reactions from the workforce. What makes this of particular interest is the potential for strong bonds between people and machines. The bond between an engineer and his train, for example, is well known and accepted. We can even observe cases where the bond exists between production workers and their machine tools: in the words of the recent Chrysler Corporation advertising campaign, "The pride is back."

We have already seen that technology infrastructure should be designed to produce quality products and services as a means of satisfying customers. And, this system should be continuously improved. We have also seen that technology systems and human workers are intimately linked; they are one and the same. They must therefore be managed as one. Thus, the workforce issue goes beyond reducing employee resistance to new technology to the question of how management prepares the workforce to adopt the philosophy that customer satisfaction through quality is the primary goal, and how technology fits into that philosophy. Technology systems that are designed without quality are bound to demoralize the workforce. But, when quality is built into the technology infrastructure from the start, then it provides workers with a clear message about the importance of quality and also allows their efforts at quality to translate into meaningful results.

Adopting the Integrated Technology Management Framework

Information technologies are so pervasive that it is difficult to imagine that any business of size would not be touched by them in some way. This does not mean, however, that these technologies will be a major component of competition in the industry or for the firm. Neither this book nor the Integrated Technology Management Framework is meant to exhort managers to adopt technologies simply for the sake of adopting them. Indeed, one of the major themes in this book has been that information technologies should be used to build long-term general competitive capacity. Yet, firms without any clear idea of how they will compete in the future or what their strategy will be will have even greater difficulty coming to grips with what the technology infrastructure should look like. We have long known that implementing computer sytems often reveals weaknesses in organizations. This means that failure to manage technology infrastructure successfully is likely to be an increasingly important indication of underlying weaknesses with management of the entire firm.

Managers should approach the Integrated Technology Management Framework as a comprehensive system. The individual components of the Framework have much less significance in isolation. Transferring knowledge about new and emerging technologies, for example, is not likely to be as successful if the workforce has not been prepared to accept and use the information. The Framework also seeks to encourage managers to undertake a major redirection in their thinking by focusing technology on quality as a means of satisfying customers, rather than on productivity improvement per se, and continuously improving the infrastructure to enhance quality further. It encourages redirection by focusing on general competitive capacity, creating a platform for short-term exploitation. This opportunistic view of strategy is different from the ends-means-ways approach American managers typically espouse. This approach to competition means that the company has bright and creative people *capable* of recognizing the opportunities and using the infrastructure's capacity to take advantage of them.

Any paradigm suffers in the general telling of it. Its usefulness rests on how well it is adapted to specific circumstances. And, that is the task that lies before American managers today.

References

"A New Era Management." *Business Week* (April 25, 1983), 50–83.

Allen, Thomas J. *Managing The Flow of Technology* (Cambridge, MA: The MIT Press, 1978).

Alter, Steven "A Taxonomy of Decision Support System." *Sloan Management Review* (Fall 1977), 39–55.

Benjamin, Robert I., Rockart, John F., Scott Morton, Michael S., and Wyman, John. "Information Technology: A Strategic Opportunity." *Sloan Management Review* (Spring 1984), 3–10.

Bolwijn, P. T., and Kumpe, T. "Toward the Factory of the Future." *The McKinsey Quarterly* (Spring 1986), 40–49.

Bowen, William. "The Puny Payoff from Office Computers." *Fortune* (May 26, 1986), 20–24.

Boynton, Andrew C., and Fmud, Robert W. "An Assessment of Critical Success Factors." *Sloan Management Review,* (Summer 1984), 17–27.

Brady, Rodney H. "Computers in Top-Level Decision Making." *Harvard Business Review* (July–August 1967), 67–76.

Browne, Malcolm W. "The Star Wars Spinoff." *The New York Times Magazine* (August 24, 1986), 18–24, 26, 66, 67, 69, 73.

Brooks, Harvey. "Technology as a factor in U.S. competitiveness." In *U.S. Competitiveness in the World Economy* edited by Bruce R. Scott and George C. Lodge (Boston: Harvard Business School Press, 1985).

Buchanan, Jack R., and Linowes, Richard. "Understanding Distributed Data Processing." *Harvard Business Review* (July–August 1980), 143–153.

Cash, James I., Jr., and Konsynski, Benn R. "IS Redraws Competitive Boundaries." *Harvard Business Review* (March–April 1985), 134–142.

Chatov, Robert. *Corporate Financial Reporting: Public or Private Control?* (New York: The Free Press, 1975).

"Computers Invade the Executive Suite." *International Management* (August 1983), 12, 13, 16–19.

Coser, Lewis *The Function of Social Conflict* (New York: The Free Press, 1956).

Curley, Kathleen Foley, and Pyburn, Philip J. "'Intellectual' Technologies: The Key to Improving White-Collar Productivity." *Sloan Management Review* (Fall 1982), 31–39.

Dammeyer, Rod F. "Developing Information Systems to Meet Top Management's Needs." *Management Review* (Spring 1983), 29, 38–39.

Daniel, D. Ronald. "Management Information Crisis." *Harvard Business Review* (September-October 1961).

Davis, David. "Computers and Top Management." *Sloan Management Review* (Spring 1984), 63–67.

Deal, Terrence and Allen Kennedy. *Corporate Cultures: The Rites and Rituals of Corporate Life* (Reading, MA.: Addison-Wesley Pub. Co., 1982).

Dearden, J. "Myth of Real-Time Management Information." *Harvard Business Review* (May–June 1966), 123–132.

Deming, W. Edwards. *Out of the Crisis.* (Massachusetts Institute of Technology, Center for Advanced Engineering Study, 1986).

Dermer, William F., Quick, Perry D., and Sandberg, Karen M. "The U.S. Trade Position in High Technology: 1980–1986." (Report prepared for the Joint Economic Committee of the United States Congress), October 1986.

Diebold, John. *Managing Information: The Challenge and the Opportunity* (New York: AMACOM, 1985).

Drucker, Peter F. "The Discipline of Innovation." *Harvard Business Review* (May–June 1985), 67–72.

Finan, William F., Quick, Perry D., and Sandberg, Karen M. "The U.S. Trade Position in High Technology: 1980–1986" (Report prepared for the Joint Economic Committee of the United States Congress), October 1986.

Fowler, Elizabeth M. "Careers: Corporate Technical Assessment." *The New York Times* (January 14, 1986).

Frude, Neil. *The Intimate Machine* (New York: New American Library, 1983).

Gannes, Stuart. "Dun & Bradstreet Deploys the Riches." *Fortune* (August 19, 1985), 38–39, 42, 46–47.

Gorry, Anthony G., and Scott Morton, Michael S. "A Framework for Management Information Systems." *Sloan Management Review* (Fall 1971), 55–70.

Guilano, Vincent E. "The Mechanization of Office Work." *Scientific American* (September 1982), 149–164.

Hayes, Robert H. "Why Strategic Planning Goes Awry." *The New York Times* (April 20, 1986).

Hunt, H. Allen, and Hunt, Timothy L. "Human Resource Implications of Robotics" (The W. E. Upjohn Institute for Employment Research, 1983).

Hurni, Melvin L. "Decision Making in the Age of Automation." *Harvard Business Review* (September-October 1955), 49-58.

"Innovation Management Practices Among Companies in North America, Europe, and Japan" (Cambridge, MA: Arthur D. Little, Inc.).

Jelinek, Mariann, and Goldhar, Joel D. "The Strategic Implications of the Factory of the Future." *Sloan Management Review* (Summer 1984), 29-36.

Johansen, Robert, and Bullen, Christine. "What to Expect from Teleconferencing." *Harvard Business Review* (March-April 1984), 4-10.

Johnson, H. Thomas, and Kaplan, Robert S. *Relevance Lost, The Rise and Fall of Management Accounting* (Boston: Harvard Business School Press, 1987).

Kanter, Rosabeth Moss. "Innovation—The Only Hope for Times Ahead?" *Sloan Management Review* (Summer 1984), 51-55.

Kaplan, Robert S. "The Evolution of Management Accounting." *The Accounting Review* (July 1984), 390-418.

Keen, Peter G. W. "Computers and Managerial Choice," *Organizational Dynamics* (Autumn, 1985), 35-49.

Keen, Peter G. W. "Decision Support Systems: Translating Analytical Techniques into Useful Tools." *Sloan Management Review* (Spring 1980), 33-44.

Kleinfield, N. R. "Turning McGraw-Hill Upside Down." *The New York Times* (February 2, 1986).

Kotter, John P. "What Effective Managers Really Do." *Harvard Business Review* (November-December 1982), 156-167.

Kotter, John P., and Schlesinger, Leonard A. "Choosing Strategies for Change." *Harvard Business Review* (March-April 1979), 106-114.

Kutter, Jeffrey. "Citibank Touts Newly Designed ATMs as a Breakthrough." *American Banker* (April 1, 1987), 8.

Laubach, Peter B. and Lawrence E. Thompson, "Electronic Computers: A Progress Report," *Harvard Business Review* (March-April 1955), 120-128.

Lawler, Edward E., and Rhode, John Grant. *Information and Control in Organizations* (Pacific Palisades, CA: Goodyear Publishing Co., 1976).

Little, John D. C. "Information Technology in Marketing" (address to the International Conference on Information Systems, December 15, 1986).

Magee, John F. "What Information Technology Has in Store for Managers." *Sloan Management Review* (Winter 1985), 45-49.

Maidique, Modesto A., and Hayes, Robert H. "The Art of High-Technology Management." *Harvard Business Review* (Winter 1984), 17-29.

McFarlan, F. Warren. "Information Technology Changes the Way You Compete." *Harvard Business Review* (May-June 1984), 98-103.

McFarlan, F. Warren. "Portfolio Approach to Information Systems," *Harvard Business Review* (September-October 1981), 142-150.

McFarlan, F. Warren, and McKenney, James L. "The Information Archipelago—Governing the New World." *Harvard Business Review* (July-August 1983), 91-99.

McFarlan, F. Warren, McKenney, James L., and Pyburn, Philip. "The Information Archipelago—Plotting a Course" *Harvard Business Review* (January-February 1983), 145-156.

McKenney, James L., and McFarlan, F. Warren. "The Information Archipelago—Maps and Bridges." *Harvard Business Review* (September-October 1982), 109-119.

McNeil, Russel B. "Mechanizing Paper Work," *Harvard Business Review* (July 1948), 492-512.

Mertes, Louis H. "Doing Your Office Over—Electronically." *Harvard Business Review* (March-April 1981), 127-135.

Miles, Raymond E., and Snow, Charles C. "Network Organizations: New Concepts for New Forms." *California Management Review* (Spring 1986), 62.

Miller, Victor E. "Decision-Oriented Information." *Datamation* (January 1984), 159-162.

Mills, Miriam K. "Teleconferencing—Managing the 'Invisible Factor.'" *Sloan Management Review* (Summer 1984), 63-65.

Monger, Rod F. "Why Seize the AI Initiative?" *Information Strategy,* (Spring 1987), 4-10.

Mooney, Marta. "Process Management Technology." *National Productivity Review* (Fall 1986), 386-391.

Noble, David F. *Forces of Production, A Social History of Industrial Automation* (New York: Alfred A. Knopf, 1984).

Nolan, Richard L. "Managing the Crisis in Data Processing." *Harvard Business Review* (March-April 1979), 115-126.

Nolan, Richard L., Brockway, Douglas W., and Tuller, Charles N. "Creating an Information Utility." *Stage by Stage,* (March-April 1986), 1-9.

Nolan, Richard L., and Gibson, Cyrus F. "Managing the Four Stages of EDP Growth." *Harvard Business Review* (January-February 1974), 76-88.

Nolan, Richard L., and Pollock, Alex J. "Organization and Architecture, or Architecture and Organization." *Stage by Stage,* (September-October 1986), 1-10.

Parsons, Gregory L. "Fitting Information Systems Technology to the Corporate Needs: The Linking Strategy" (Boston: Harvard Business School, 1983).

Parsons, Gregory L. "Information Technology: A New Competitive Weapon." *Sloan Management Review* (Fall 1983), 3-14.

Perrow, Charles. *Complex Organizations, A Critical Essay* (New York: Random House, 1986).

Poppel, Harvey L. "Who Needs the Office of the Future." *Harvard Business Review* (November-December 1982), 146-155.

Porter, Michael E. *Competitive Advantage, Superior Performance* (New York: The Free Press, 1985).

Porter, Alan L., Rossine, Frederick A., Carpenter, Stanley R., Roper, A. T., Tiller, Rolan W. *A Guidance for Technology Assessment and Impact Analysis* (New York: North Holland, 1980).

President's Commission on Industrial Competitiveness. *Global Competition, The New Reality* (Washington, D.C.: Superintendent of Documents, U.S. Government Publications Office, 1985).

Press, Frank. "Technological Competition and the Western Alliance." In *A High Technology Gap? Europe, America, and Japan* (New York: Council on Foreign Relations, 1987).

Quinn, James Brian. "Technological Innovation, Entrepreneurship, and Strategy." *Sloan Management Review* (Spring 1979), 19–30.

Roberts, Edward B., and Alan R. Fusfeld. "Staffing the Innovative Technology-Based Organization." *Sloan Management Review* (Spring 1981), 19–34.

Rockart, John F. "The Changing Role of the Information Systems Executive: A Critical Success Factors Perspective." *Sloan Management Review,* vol. 24, no. 1 (Fall 1982), 3–13.

Rockart, John F., and Crescenzi, Adam D. "Engaging Top Management in Information Technology." *Sloan Management Review* (Summer 1984), 3–16.

Rockart, J. F. "Chief Executives Define Their Own Data Needs." *Harvard Business Review* (March–April 1979), 81–98.

Sales, Amy, and Philip, Mervis. "The Impact of New Technology on People in Organizations: A Review of the Current Literature." (Unpublished paper, 1985).

Servan-Schreiber, J. J. *The American Challenge* (New York: Atheneum, 1967).

Shaiken, Harley. *Work Transformed* (Lexington, MA: D. C. Heath and Company, 1984).

Slauterback, William H., and Werther, William B., Jr. "The Third Revolution: Computer-Integrated Manufacturing." *National Productivity Review,* (Autumn 1984), 367–374.

Smilor, Raymond W., and Gill, Michael D., Jr. *The New Business Incubator* (Lexington, MA: Lexington Books, 1986).

Sommers, Albert T. *Economic Policy Needs for the Coming Decade* (New York: The Conference Board, Inc., 1983).

Taylor, Arthur R. "Long Live Marketing." Working Paper, Fordham Business School, 1986.

"The Forces Behind White-Collar Layoffs." *New York Times* (October 13, 1985).

"The Information Business." *Business Week* (August 25, 1986), 82–86, 90.

Waterman, Robert H., Jr., Peters, Thomas J., and Phillips, Julien R. "Structure Is Not Organization." *Business Horizons,* (June 1980), 14–26.

Zuboff, Shoshana. "Automate/Informate: The Two Faces of Intelligent Technology." *Organizational Dynamics,* (Autumn 1985), 4–18.

Index